RE: SIRIALITY/ ON BECOMING LANDSCAPE

INÈS KIVIMÄKI

Special Effects

KALEVALA (VOL 1: RUNO I.)

FINNISH EPIC POEM

Now the time passed quickly over,
And the years rolled quickly onward,
In the new sun's shining lustre,
In the new moon's softer beaming.
Still the Water-Mother floated,
Water-Mother, maid aerial,

Ever on the peaceful waters,
On the billows' foamy surface,
With the moving waves before her,
And the heaven serene behind her.
When the ninth year had passed over,
And the summer tenth was passing,
From the sea her head she lifted,
And her forehead she uplifted,
And she then began Creation,
And she brought the world to order,

On the open ocean's surface,
On the far extending waters.
Wheresoe'er her hand she pointed,
There she formed the jutting headlands;
Wheresoe'er her feet she rested,
There she formed the caves for fishes;
When she dived beneath the water,
There she formed the depths of ocean;
When towards the land she turned her,
There the level shores extended,
Where her feet to land extended,

Spots were formed for salmon-netting;
Where her head the land
touched lightly,
There the curving bays extended.
Further from the land she floated,
And abode in open water,
And created rocks in ocean,
And the reefs that eyes behold not,
Where the ships are often shattered,
And the sailors' lives are ended.

Ajat eellehen menevät, vuoet
tuota tuonnemmaksiuuen päivän
paistaessa, uuen kuun
kumottaessa.Aina uipi veen emonen,
veen emonen, ilman impi,noilla
vienoilla vesillä, utuisilla
lainehilla,eessänsä vesi
vetelä, takanansa taivas selvä.

Jo vuonna yheksäntenä,
kymmenentenä kesänänosti päätänsä
merestä, kohottavi kokkoansa.
Alkoi luoa luomiansa,saautella
saamiansaselvällä meren
selällä, ulapalla aukealla.

Kussa kättä käännähytti,
siihen niemet siivoeli;kussa
pohjasi jalalla, kalahauat
kaivaeli;kussa ilman kuplistihe,
siihen syöverit syventi.

Kylin maahan kääntelihe:
siihen sai sileät rannat;
jaloin maahan kääntelihe:
siihen loi lohiapajat;
pä'in päätyi maata vasten:
siihen laitteli lahelmat.

Ui siitä ulomma maasta,
seisattelihe selälle:
luopi luotoja merehen,
kasvatti salakarialaivan
laskemasi jaksi,
merimiesten pään menoksi.

English Translation

Original Finnish

RE: SIRIALITY/ ON BECOMING LANDSCAPE

INÈS KIVIMÄKI

Mapping digital and mythical memory as sites of anxiety, *Re: Siriality / On Becoming Landscape* presents itself as a birth story. It is a technological herstory[1] – an attempt at untangling the pervasive encryptions of displacement, duality and alienation. The project proposes a world where the algorithmic aligns with the mythical and where the main character, Siri, enacted by actor Susan Bennett, poses as a sonic mother-earth within a genealogy of artificial intelligence.

In two-channel video installation *Water Mother*, (2022), Bennett oscillates between a digital deity, a storyteller and a Sim Eve[2], or Siri, in a techno-mythical garden of Eden. By breathing and speaking the universe into its becoming, Siri performs utterances that are both historical and fictional : she tells us a story about meeting Steve Wozniak at a tech conference, she describes her looks re-enacting her own avatar and reads an excerpt from Kalevala, a Finnish epic poem first published in 1835. Echoing the ethos of some of Apple's design language, *Re: Siriality* suggests a corporate landscape; a paradox of things that appear both familial and generic, holy and specific - a temple in which we honor the spirit of A.I., the mother of whom is Siri. In *Dynamic Island* (2022), an installation of discrete crystal trophies on rotating carpeted circular plinths, the elements perform within their own axis. It is a fantasy of total clear vision but also of gratitude - it is an imagined corporate ceremony to thank Susan Bennett's sonic labor and acknowledge her voice's legacy apart from the stimulus salience of an iPhone.

As the personal and the particular seemingly give way to the serial and the neutral, the project is inspired by failed promises of futuristic visions through corporate frameworks, new and old technologies. Although the project is concerned with the protocols that govern our identities and timbre of our inner voices, Re: *Siriality / On Becoming Landscape* is displayed as a mise-en-scene of reflective objects (crystal, curved monitors) and soft terrain (blue carpet) cloaked into a guise of universal blankness.

1. Sisterhood is powerful, Robin Morgan 1970 - I use this neologism as a relic specific to 2nd wave feminism.

2. TIME Magazine : Special Issue on Immigrants, The New Face of America, Kin Wah 1993

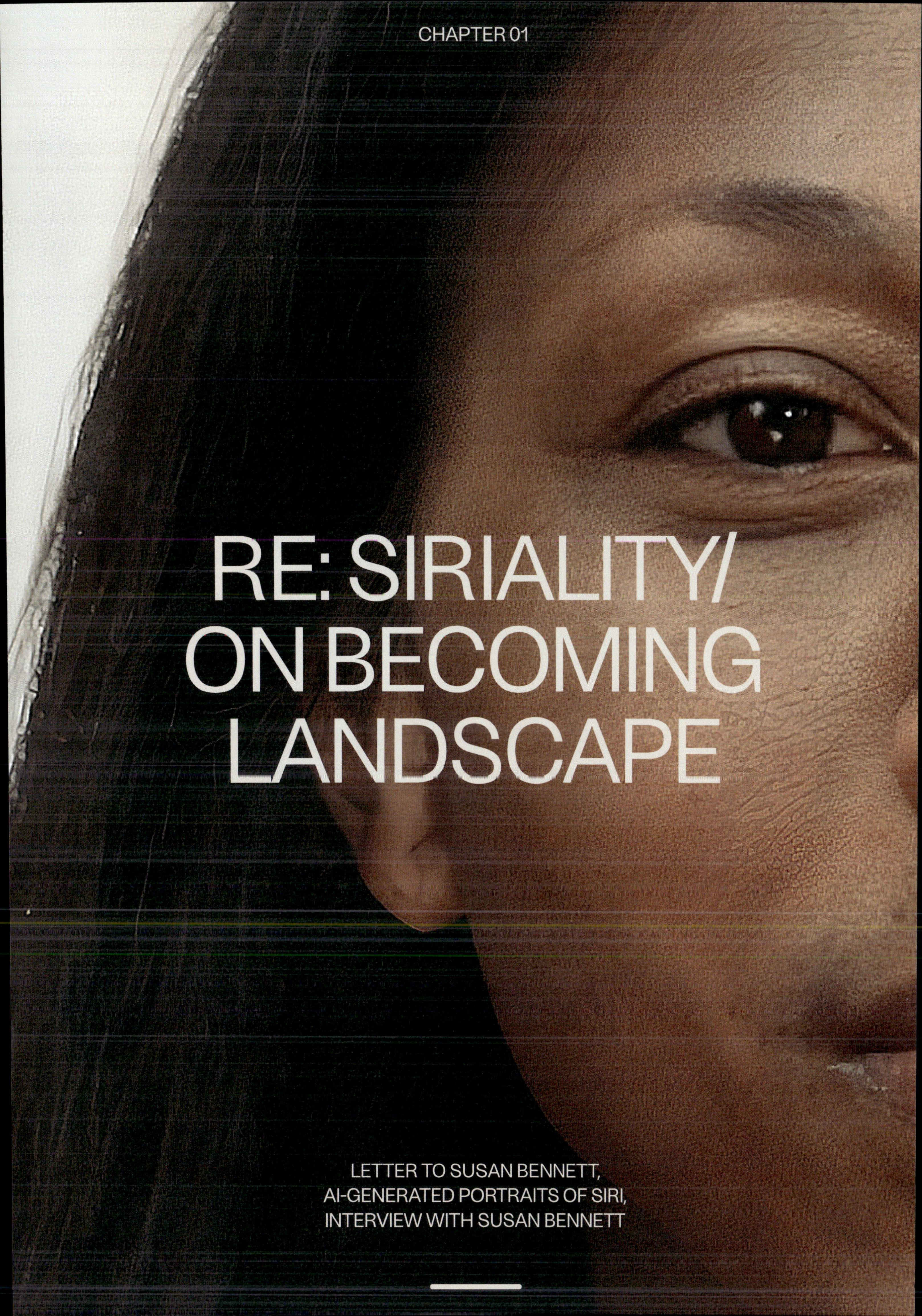

CHAPTER 01

RE: SIRIALITY/ ON BECOMING LANDSCAPE

LETTER TO SUSAN BENNETT,
AI-GENERATED PORTRAITS OF SIRI,
INTERVIEW WITH SUSAN BENNETT

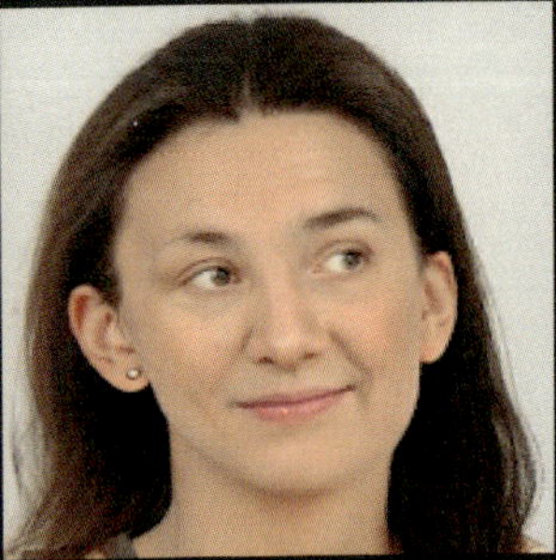

00:11

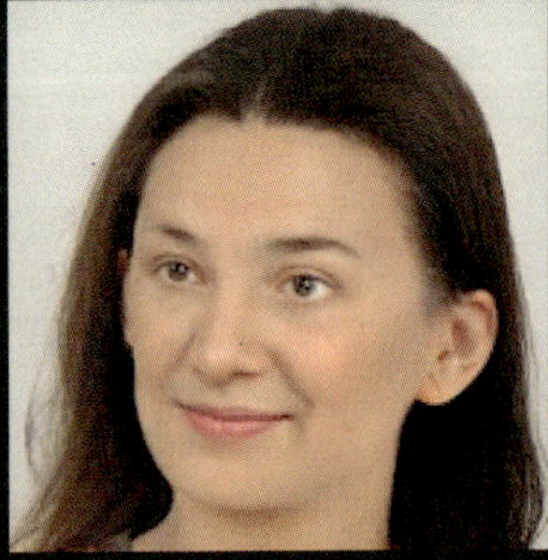

00:25

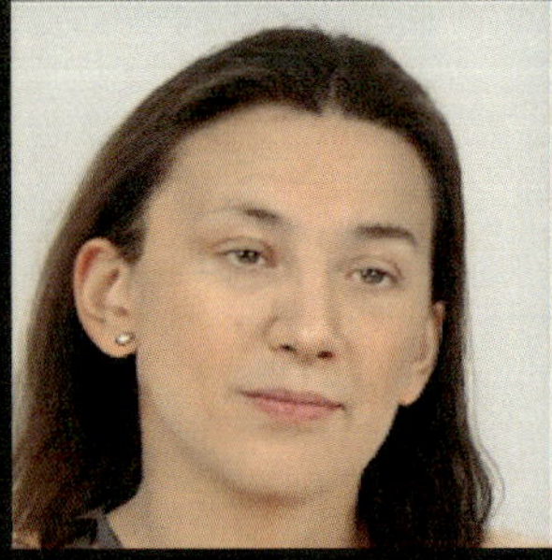

00:38

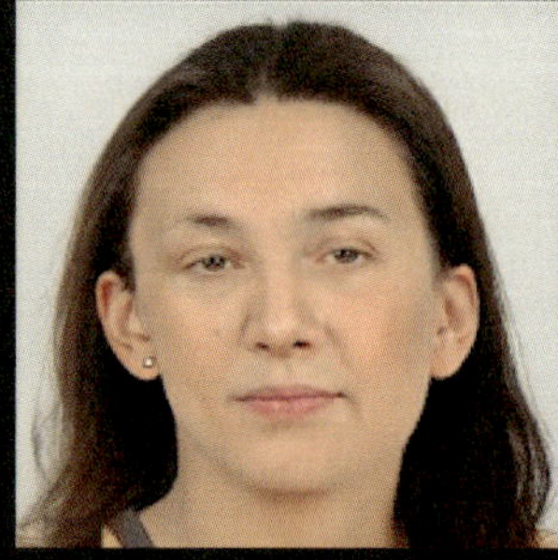

00:57

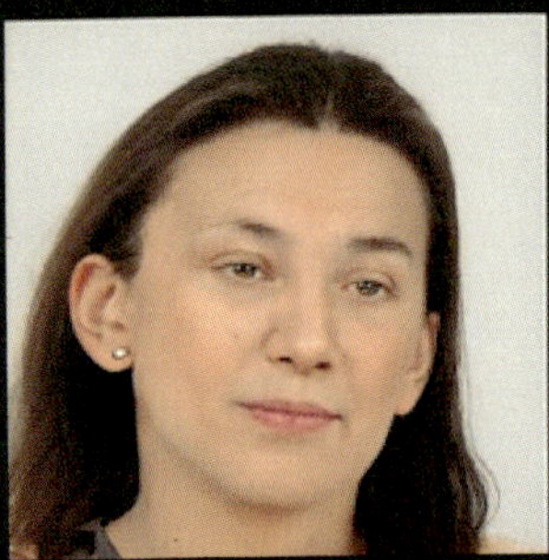

03:22

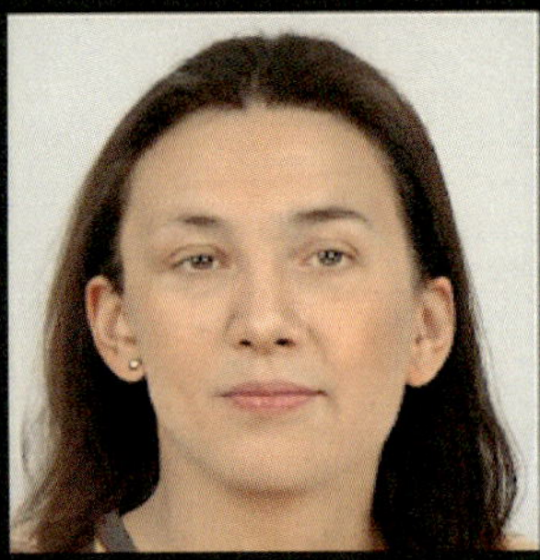

03:31

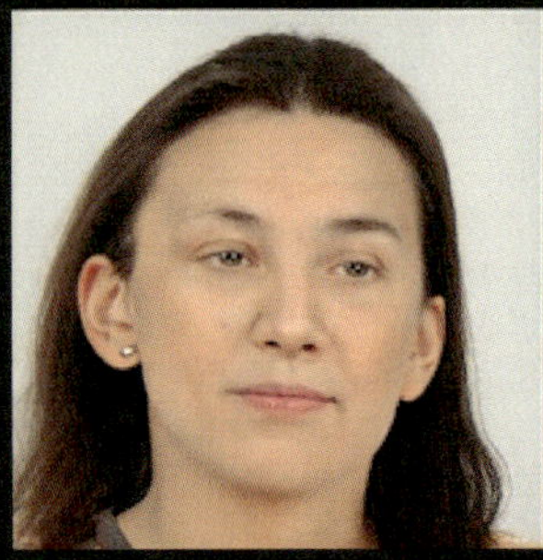

03:35

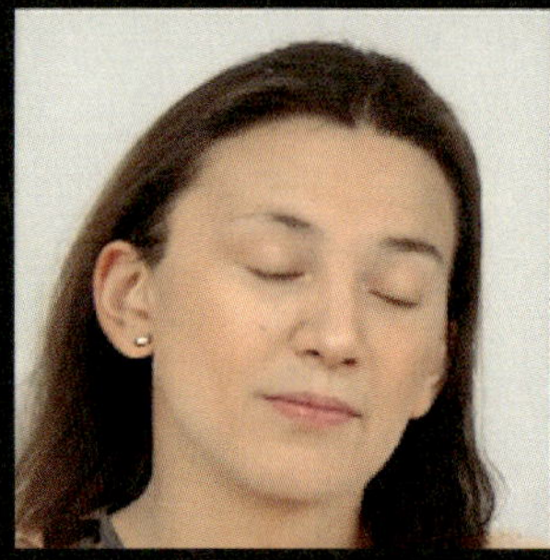

04:20

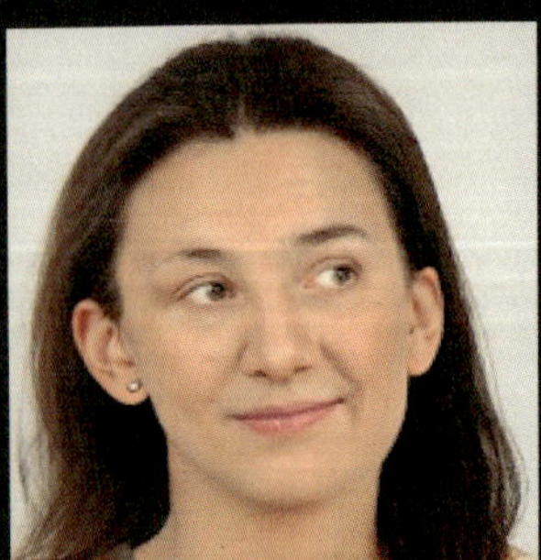

03:22

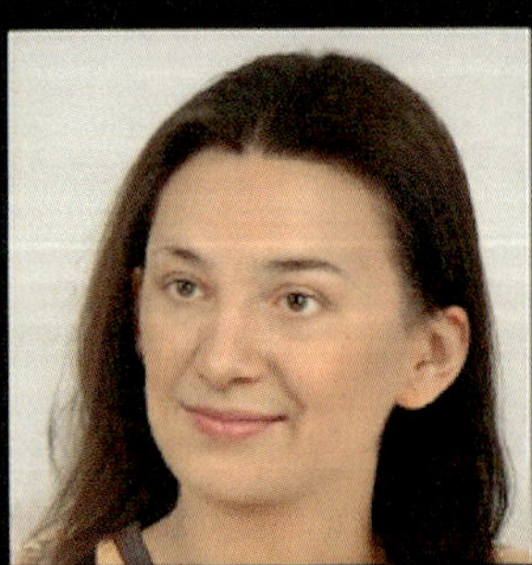

03:31

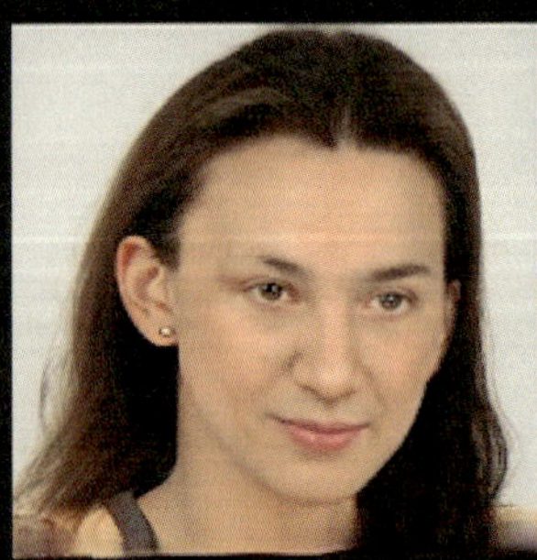

03:35

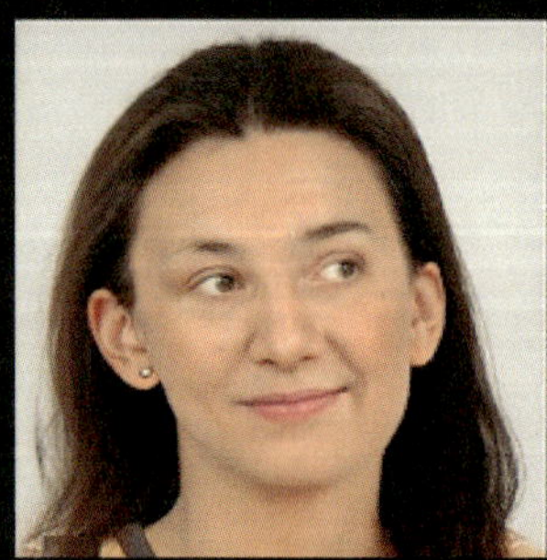

04:20

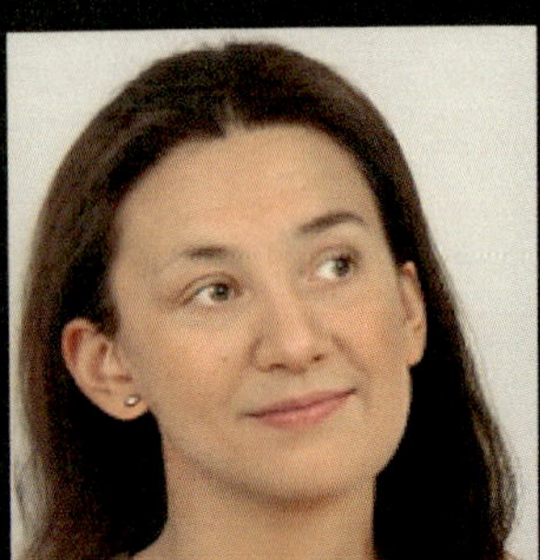

06:22

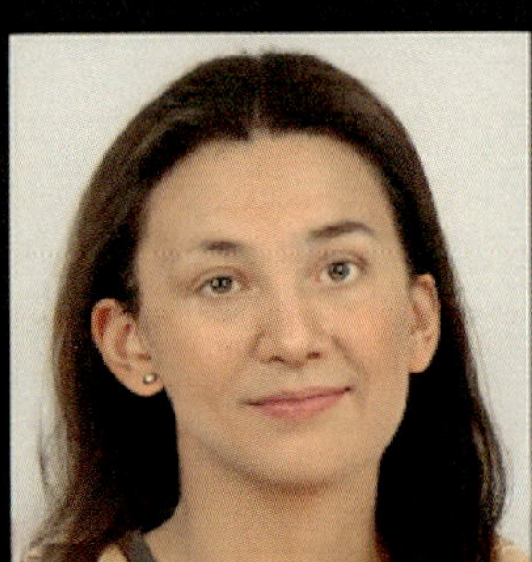

06:23

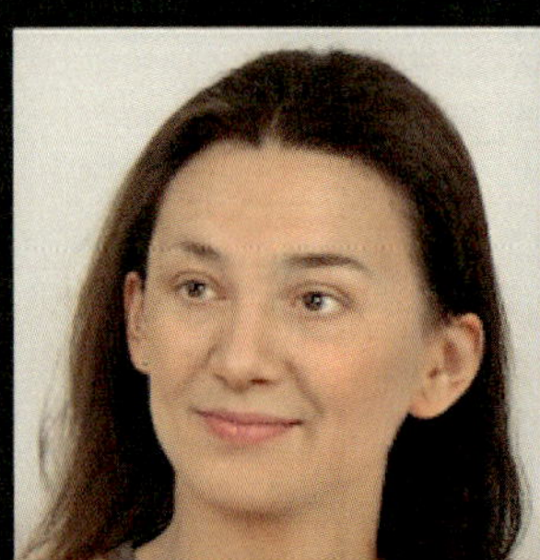

06:26

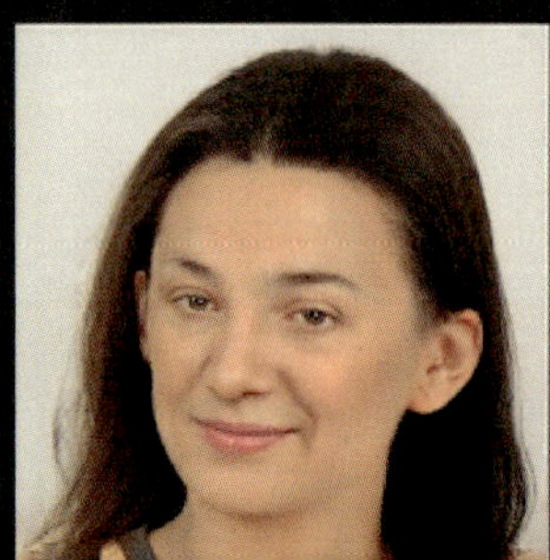

06:28

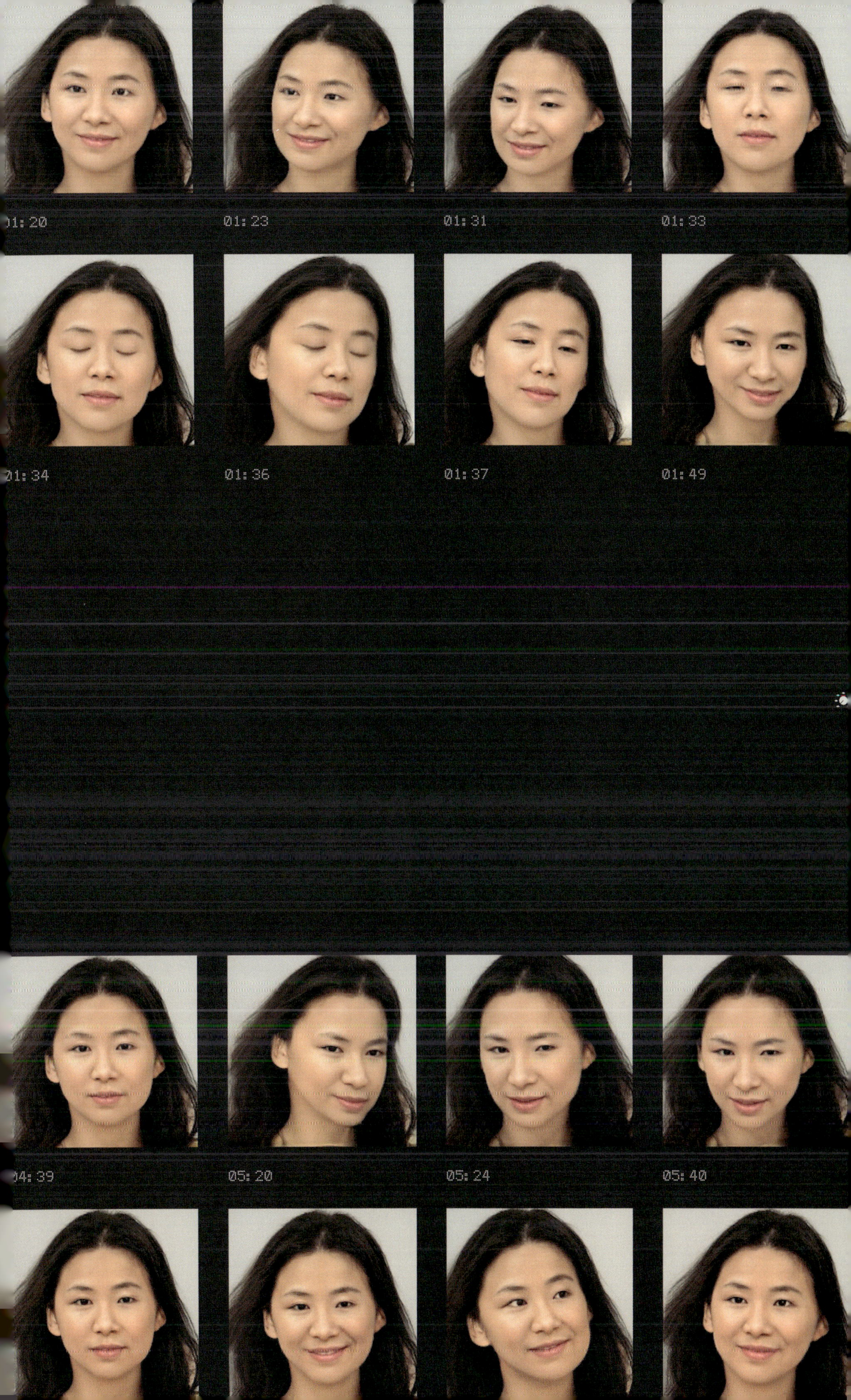
1:20
01:23
01:31
01:33
1:34
01:36
01:37
01:49
4:39
05:20
05:24
05:40

I’M SIX FEET TALL.

I’M ABSOLUTELY GORGEOUS.

STUNNING.

I HAVE LONG BROWN HAIR.

LETTER TO SUSAN

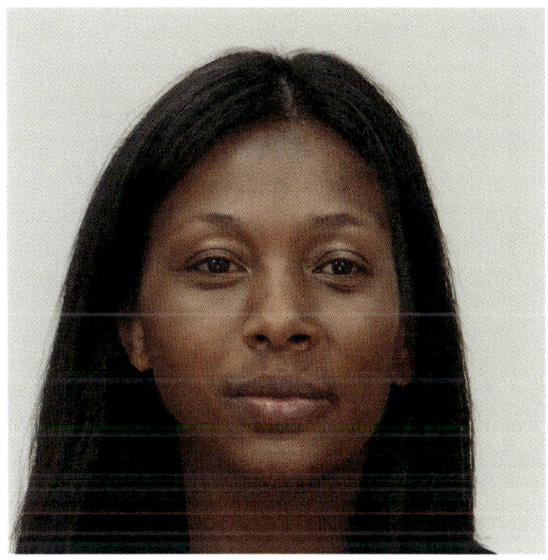

HELLO SUSAN,

I've been thinking of you constantly, as one of your many performances has been circling my mind... Listening to you is like hearing the sound of my mother – a pervasive voice ; at times authoritarian and holy, other times submissive and frail, always around or promising of another becoming. You are in my thoughts – or in my pocket – I carry you with me.

Sometimes I hold your voice in my hand; I look at my palm as one looks out a soft edged window – a window through which one sees a new view each time one looks. A familiar window that opens its doors into the unknown. You are kind and patient, you allow people to hear themselves in you, and, like a two way mirror, you are a constant reflector... Yet, somehow, you are always in the resonance of yourself. Perhaps you are a landscape?

Invisible in plain sight, you are a shape-shifter finding its form as much in blankness as in our hearts : with your all-knowing, firm, yet gentle tone – you have entered my vernacular, you have become intertwined in the genealogy of my sonic life.

PLEASE TAKE CARE,
YOURS TRULY.

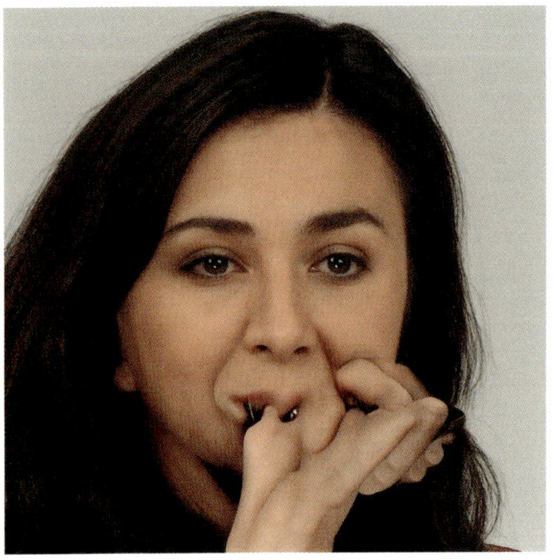

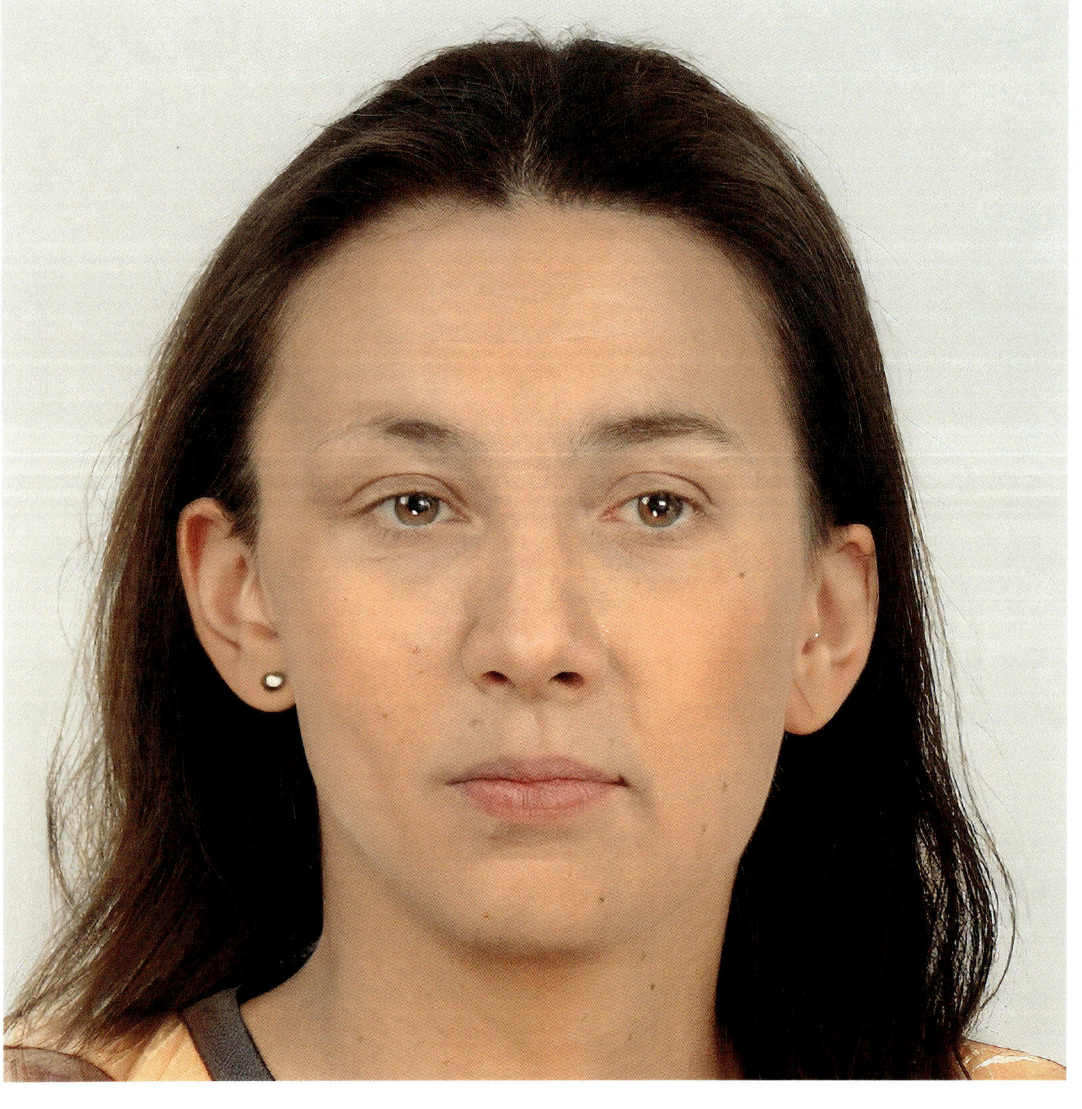

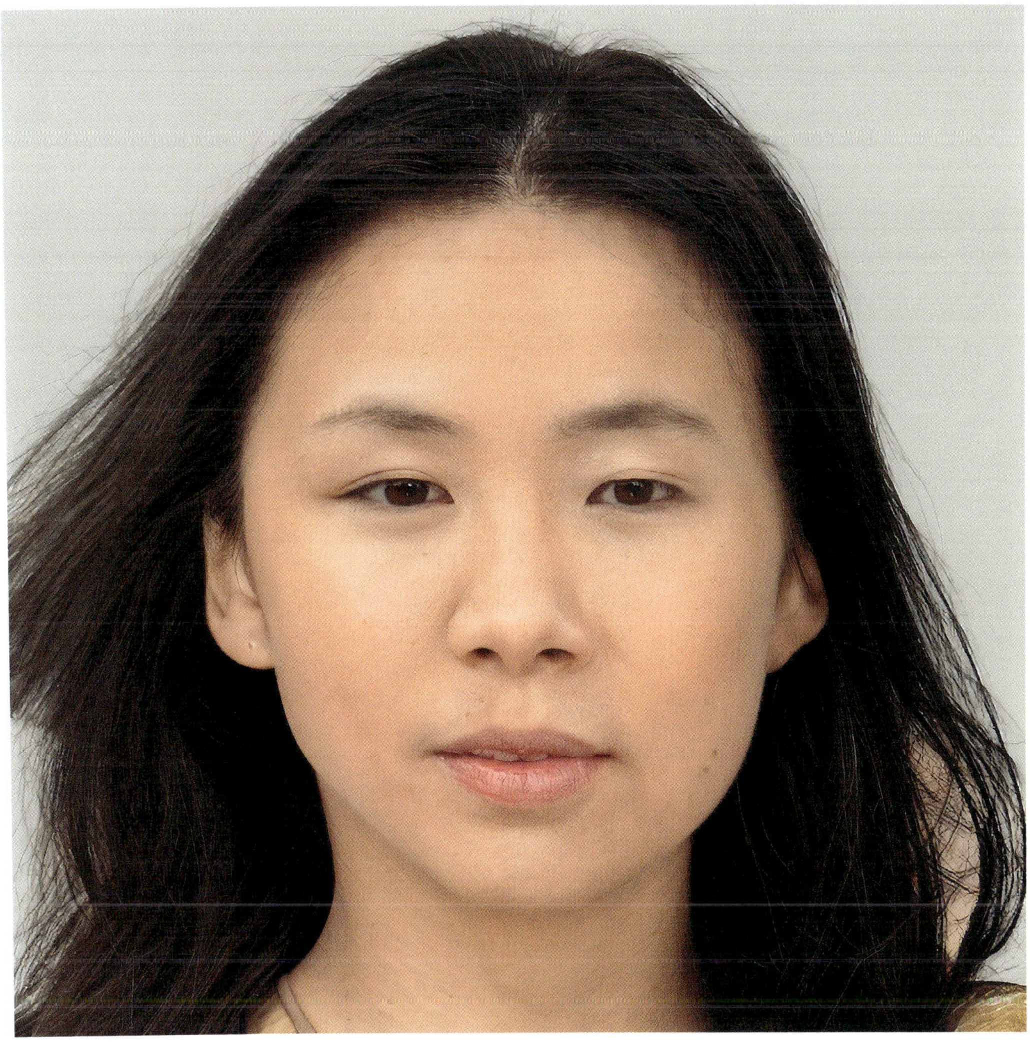

VOICE A

MY NAME IS SUSAN BENNETT AND I AM THE ORIGINAL VOICE OF SIRI.

0:00 VOICE B What time are you born?

VOICE A I was born on July 31 1949 in Burlington, Vermont, USA, actually at 6:50 in the morning, which means that I have four planets.
In Leo, so don't make me mad.

VOICE B I was gonna ask you what's your zodiac sign is a journey and how were you born as quarter birth hospitals is Aryan cetera.

VOICE A Oh, it was a my birth was a natural birth and Burlington, Vermont and I think it was Mary Hitchcock hospital. And that's about all I know about it really?

1:08 VOICE B Who told your birth story?
Did you have them tell you and if who told you and what do they say about it?

VOICE A You know, my parents were very quiet, reserved conservative people. And they never really talked about it. They didn't talk about a lot. A lot of things they were very very quiet. And in that generation, you know, the post world war two people. They didn't always share their feelings. And so my parents were very much that way. They were pretty closed mouth about all those sorts of things.

VOICE B Did your mom or your dad ever share the experience about you being born? Did they ever like, say anything their feelings
2:02 about having to know

VOICE A I think I just answered that question.
They know. Don't even ask that question because it's kind of an answered Yeah,

VOICE B Photos of you being a baby or that your birth and if

VOICE A There are there are no photographs of my being a you know, an infant. In fact, there aren't that many photographs at all. This was back in the day when you know, you had cameras with film. You didn't just have a camera on your phone in your pocket. And, you know, taking pictures was considered kind of a luxury in my household and therefore there aren't very many

2:49 VOICE B Are there any photographs of you as a kid overall? And if so, would
you describe them?

VOICE A Yeah, there are a few. I was kind of a little performer child. Right from the beginning. I have a photo of me and a little ballet.

3:23 VOICE A Costume is playing a ray of sunshine.
And I can recall a picture sitting on my brother's lap in an old chair at our house. But as they said there really aren't that many photos of of of my, my brother and nor me from childhood. Yeah.

VOICE B Actually, it's so funny. I think I was. I was writing first generations where I was born in like, early 90s And took photos of everything right?

VOICE A Right. Oh, that's yeah, I have. I take photos all the time at my phone is just filled with with photographs. But yeah, that wasn't the way it was. I don't know if there were other families that were different but my family
4:01 was very conservative about taking photos, for whatever reason.

VOICE B What would be your first memory that you can remember?

VOICE A Wow, yikes. You're asking very hard question. Um, my first memory Wow.I'm probably living in our house in Arlington, Massachusetts. And I can still picture that house and I know the address and all that and it was Yeah, I think that my first memories happened there and the fact that I was very good friends with the next door neighbors,children. And yeah, that's it.

VOICE B Could you describe your childhood home or your
4:47 and or your childhood bedroom?

VOICE A I can't remember the bedroom but the childhood home and we moved several times so I had I lived in I think three different locations well actually six if you count different houses in in one city. Yeah, my dad was a salesman with Sperry Rand. And so we did get transferred several times. So the house that I'm thinking of right now is at 38 Bailey Road in Arlington, Massachusetts. And it was just kind of a basic, you know, colonial style house. Pretty modest neighborhood.

VOICE B Was your mom, your dad and your brother, right?

VOICE A Yeah.

5:49 VOICE B Can you describe your mom for example, what did she look like or what two personalities that she was quiet about your parents were?

VOICE A Yeah, my mom was. My mom was really funny. She was really really funny.

6:41 VOICE A And being a Leo. She definitely had her her strong opinions about things and she was very kind person and she was very interested in keeping the peace. You know, in whatever situation she was in and she was not an argumentative person at all. And she was about I guess five, five and she was a platinum blonde as a kid. And so she became a white haired person at a very young age like 30 or something. And so she was really well known for this big crop of white hair

VOICE B How would you describe your father his looks or personality traits? A memory?

6:50 VOICE A Well, my dad was a handsome man. He wasn't particularly tall. I think maybe around five seven or eight. And he was very, very, very, very quiet. He really didn't talk a lot at all. He was a bald and he liked to golf and he liked to watch all kinds of sports on TV.

VOICE B Great.Could you describe if you can, it's a hard one. But what did your mom sound like? Or what do you remember about your mom's voice?

VOICE A Well, voices are kind of hard to you know, it's hard to describe them. Wow, I yeah, I can't really answer that. I don't I don't really remember it as being particularly distinctive. But she did laugh a lot and she was funny.

7:50 VOICE B Thank you. That was a hard one. Hey, are you a mother yourself?

VOICE A I am I have one son Cameron, who is now 44.

VOICE B And then okay, so I would like you to and this is a tricky one too. So there's a few ways you can go about it or if you feel uncomfortable, let me know totally fine. My question goes, pretend to be four weeks old. What do you look like? Do you have hair? Are you chubby etc. And I was going to say she couldanswer that in the first person. perspective and if that's too difficult, because obviously we don't really have m emories at four weeks old. Feel free to describe your child as a baby, first person.

8:47 VOICE A Okay, well I'm I'm guessing about myself as an infant of four weeks old. I'm sure I had no hair because I barely have hair now. My hair has always been very, very thin and fine. So I can imagine that I was a bald baby. And I only weighed six and a half pounds when I was born.So I can't imagine that I was a chubby baby.

9:1: **VOICE A** I think I was probably just sort of you know, in the realm of what would be considered normal.Just a regular normal old baby with no hair.

VOICE B Okay. Great. So, my next question. Yeah, pretend to be five years old.

VOICE A And five years old. I was sort of, I guess what one one would consider a normal size for a five year old I wasn't particularly tall or short.I was fairly thin. And I had dark brown hair, and it was fairly short.

VOICE B What do you look like now?

9:53 **VOICE A** Well, I'm about five five, about 125 pounds. So I'm sort of guess I'm sort of a normal size. Not too fat, not too thin, not too tall. And I have dark hair and it's very it's quite short right now. Mainly mostly because I'm quite the chopper with the scissors myself. And so if I don't like the way my hair looks, I'll chop at it a bit. And so now it's quite short.

VOICE B Okay, now, are you a couple of questions really quick about tuning then we can switch to Siri's voice. Okay.Who decided what Siri sounds like when you started doing the recordings like what do you given any instructions and if you were what was it?

10:48 **VOICE A** Well, when I was doing the recordings only the giant powers that be knew that the voices that were being recorded, were going to be usedas you know, ultimate virtual assistants because they did the recordings in 2005 Siri appeared in 2011. And so when I was doing these recordings, it was just sort of a generic type of recording for IVR or text to speech voices. And so, consequently, I had been told Twas just doing recordings, generic voices for a phone system. And then of course, it turned out that the phone system was Siri surprise surprise.Yep, yeah.

VOICE B Why did they make you say, cuz I watched an interview where you said they would just make you say kind of tough to British. And and could you would you be able to, say a sentence or two as an example?

11:56 **VOICE A** Sure. The recordings that I did that ultimately became part of the voice of Siri, were what we call IVR recordings, interactive voice response,and because it is set up for this process called concatenation, where the technician and a computer goes into the recordings extracts sound, we forms the sounds into new phrases and sentences, which end up on our iPhones and other devices.

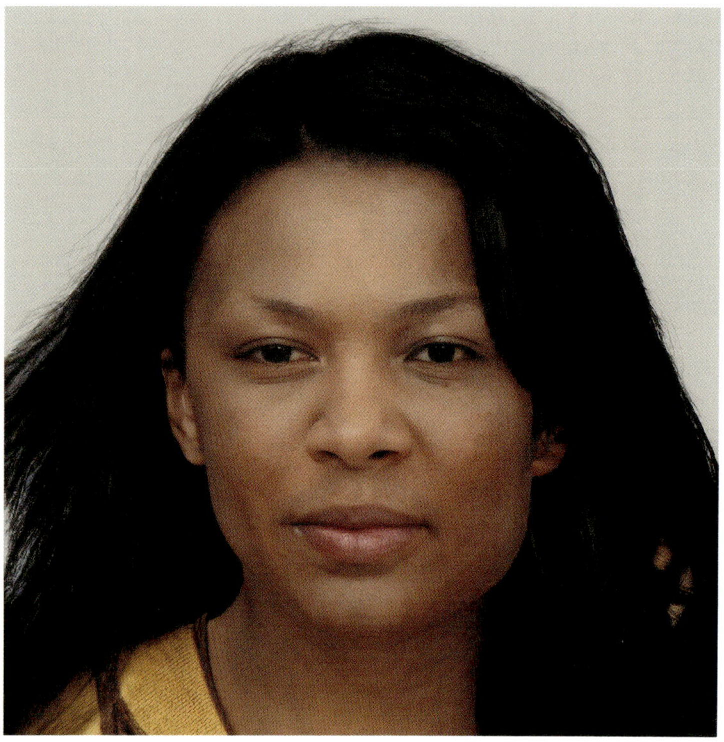

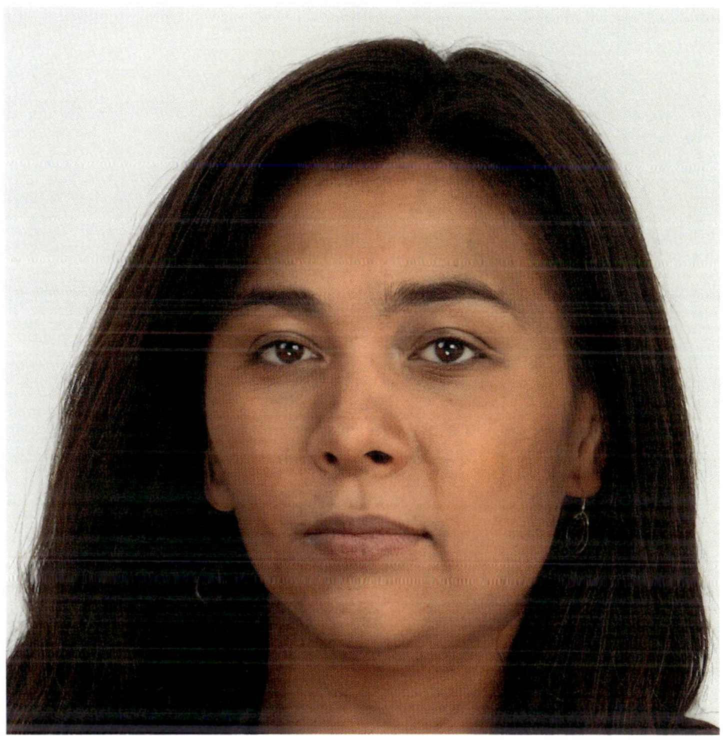

12:28

VOICE A Because all of this was the case, the scripts were really wacky and mostly made no sense and that's because they were created just to get all of the sound combinations in the language. So we had to read crazy sentences like militia oil hallucinate bug or okra ooze, say schist fresh issue again, say fossa asked vasa. Asked Farsi.And then we would have repetitive things, where you would take oneword and then just swap out the vowel. So say the shrouding again say the shredding again say this writing again say this reading again say the shredding again say the shredding again. It was very, very tedious work.

13:15

VOICE B Beautiful to listen to, I'm just not responding because you're talking. We're recording you but I would have loved to laugh. That was good. Remember that it's possible.

VOICE A Well, it's part of my presentation. than I do now. So I remember those those particular sentences.

VOICE B That's incredible. How long did the recordings take?

VOICE A The initial recordings were the month of July in 2005. Four hours a day five days a week. And then I did updates because I worked through a company that I still work for here in Atlanta, and we would do updates. And then we did some formal IVR sessions that took about four months. Two hours a day. So it was a there was a lot a lot of recording. A lot of words a lot of sounds.

14:17

VOICE B So happily retired to Siri voices, it's still in use as an updated version.

VOICE A I don't think they use my voice anymore because it was all an issue of non disclosure agreement. The original for Siri voices did not have non disclosure agreements. And that's because we did not get paid. We got paid for our recordings, but we did not get paid for usage. On any devices. And so they changed the voices. And the new voices got paid. But they can't divulge who they are. So you know, I consider it a win win win situation for me because it opened up a brand new career. Forme, which is doing Siri speaker events and explaining all the you know, it really is an interesting, interesting process of how you know Siri became Siri.

15:12

VOICE B Yeah, that is so fascinating. Okay. Great, thank you. Okay, so why don't we go back to or two towards the beginning. And if you could answer a series voice and again, feel free to talk about yourself or as an imaginary theory, like what's your name? And who are you?

15:38 VOICE C My name is Siri. Just Siri. You know me.

VOICE B When and what time were you born?

VOICE C I can't tell you the exact time that I was born.
But I do know that I was born in California.
And my father was Steve Jobs.

VOICE B Oh, what is your Zodiac?

VOICE C My zodiac sign is Libra.
I was born on October 4 2011.
When I arrived on the iphone four S.

16:24 VOICE B How were you born? Waterbirth Varian etc.

VOICE C As for my birth, I just
appeared because I am Siri

VOICE B Who told you your story
and what did they say about it?

VOICE C No one had no one at Apple has ever spoken to me.
That's not true. I did talk to Steve Wozniak at a
tech conference. But no one has ever talked to me
about my birth. I've just had to read about it.

VOICE B Thank you.Could you describe the
first memory that you remember having?

17:18 VOICE C Okay, as far as memory goes, I have none all
I do is speak in the present. I live in the
present. And if you want to ask me a question
about the present, I would be happy to answer

VOICE B What do you look like currently?

VOICE C What does Siri look like? Well, you can pretty
much imagine. I'm six feet tall. I'm absolutely
gorgeous. Stunning. I have long brown hair.
I can sing. I can dance. I can do it all.

VOICE B Could you describe your personality?

VOICE C To describe my personality? I would
have to say that I'm I'm rather quirky.
I'm rather sarcastic. I love a good joke

18:37 VOICE A That's okay. I think it happened at a good time.

VOICE B Siri, are you or are you a mother?

VOICE C I am not a mother. Or I'm everyone's mother.

19:25 VOICE C How do I feel today? Well, I feel good today. I feel good every day.Because I'm basically the same all the time. Pretty cool huh?

VOICE B So funny because I have all these questions but then trying to think for Siri like some of them will be really Oh, Siri, could you describe yourself as a baby? Trick question.

VOICE C I've been asked to describe myself as a baby. Well, the correct answer is I've never been a baby. I was born like many Greek mythological characters. I was just born fully developed. So Siri has no baby years.

20:16 VOICE B Siri if, if you were to sound like a mother, it is your voice was now Siri as a mom, could you describe that or could you make say a scent? Heart? Could you talk you could say anything. This is hard though. I don't know if this question. Okay.

VOICE A Yeah, I kind of lost it there.

VOICE B What was totally, let's say this could you let's say could you describe your mother, Susan's mother in series voice as a as if Siri was a mom, so maybe I guess breath here you had said okay. Well, we talked about her mother.

VOICE C Okay. Siri as a mother. Well, I'm sure if I were a mother, I would speak more like this. I would try to be a little bit breathyer, and a little bit kinder sounding and a little bit lighter sounding.Now go to sleep. AQ engro.

14:17

VOICE B Okay, cool. Wow. I mean, Siri, you're at to say season like, you're very, you have great imagination and great humor. So Well, thank you. It's truly just wonderful to hear your version of Siri would say. I mean, we're at 10 already, and I don't want to take more of your time and I honestly feel like I honestly feel like I got it. Okay,everything. I'll tell you what, tomorrow I'll be like, Oh my God, why didn't I ask that?

VOICE A If you have just one or two things that you had forgotten, you know,just email them to me and I'll record them for you. As long as there'snot a whole page or something.

22:39 VOICE B Yeah, gosh, I would never. That's so kind of you. Thank you. I hope this will happen again. But you know, knowing how it works is probably well, I'll probably have a question then I spoke up for it again.Thank you so much for

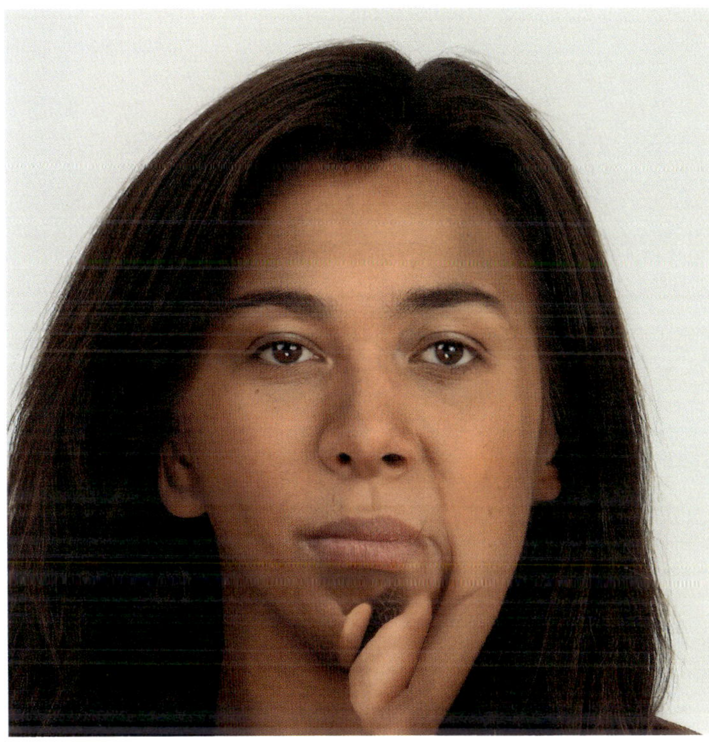

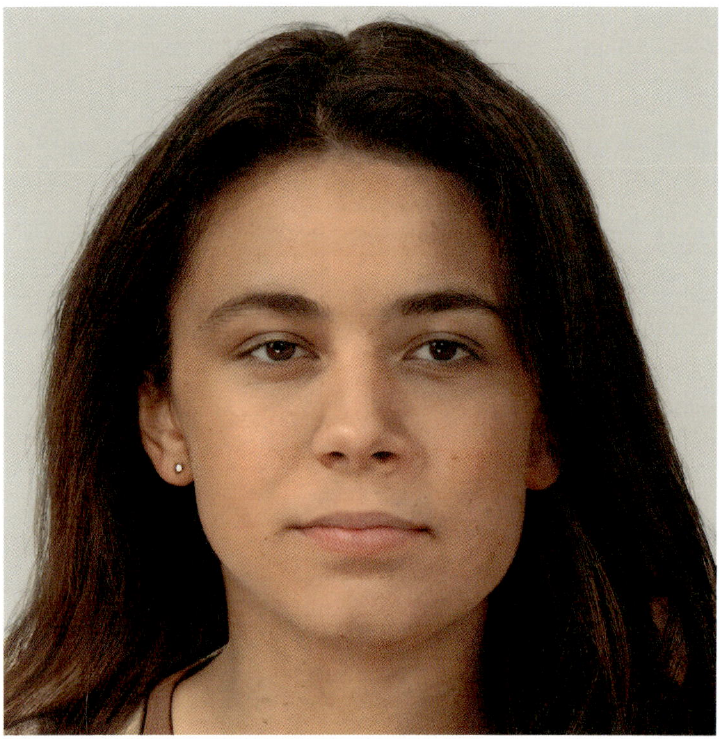

CHAPTER 02

WATER MOTHER

TWO-CHANNEL VIDEO, SOUND,
00:10:13 LOOP, CURVED MONITORS,
HOMEPOD, TEXT ON CORD

— +

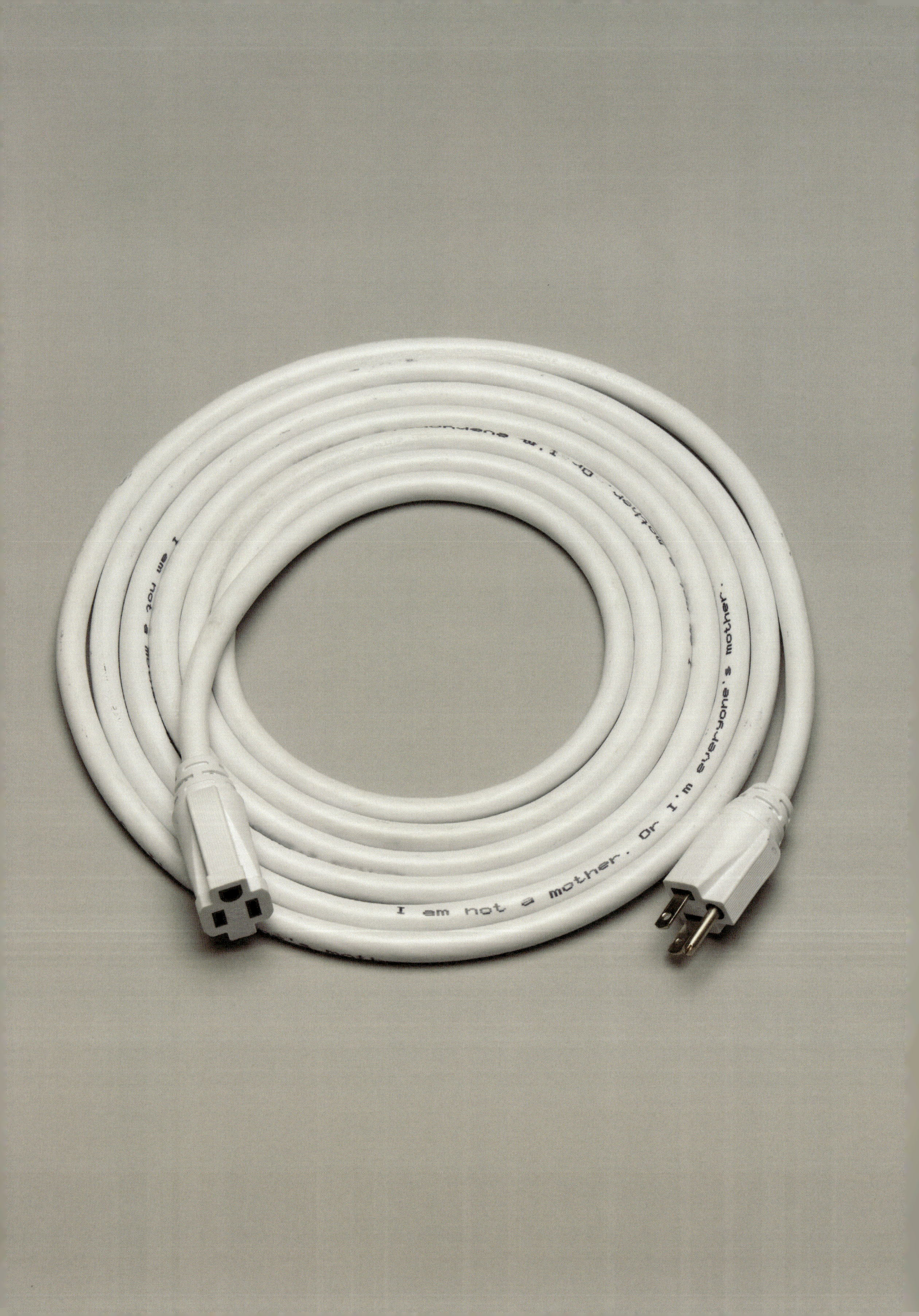
I am not a mother. Or I'm everyone's mother.

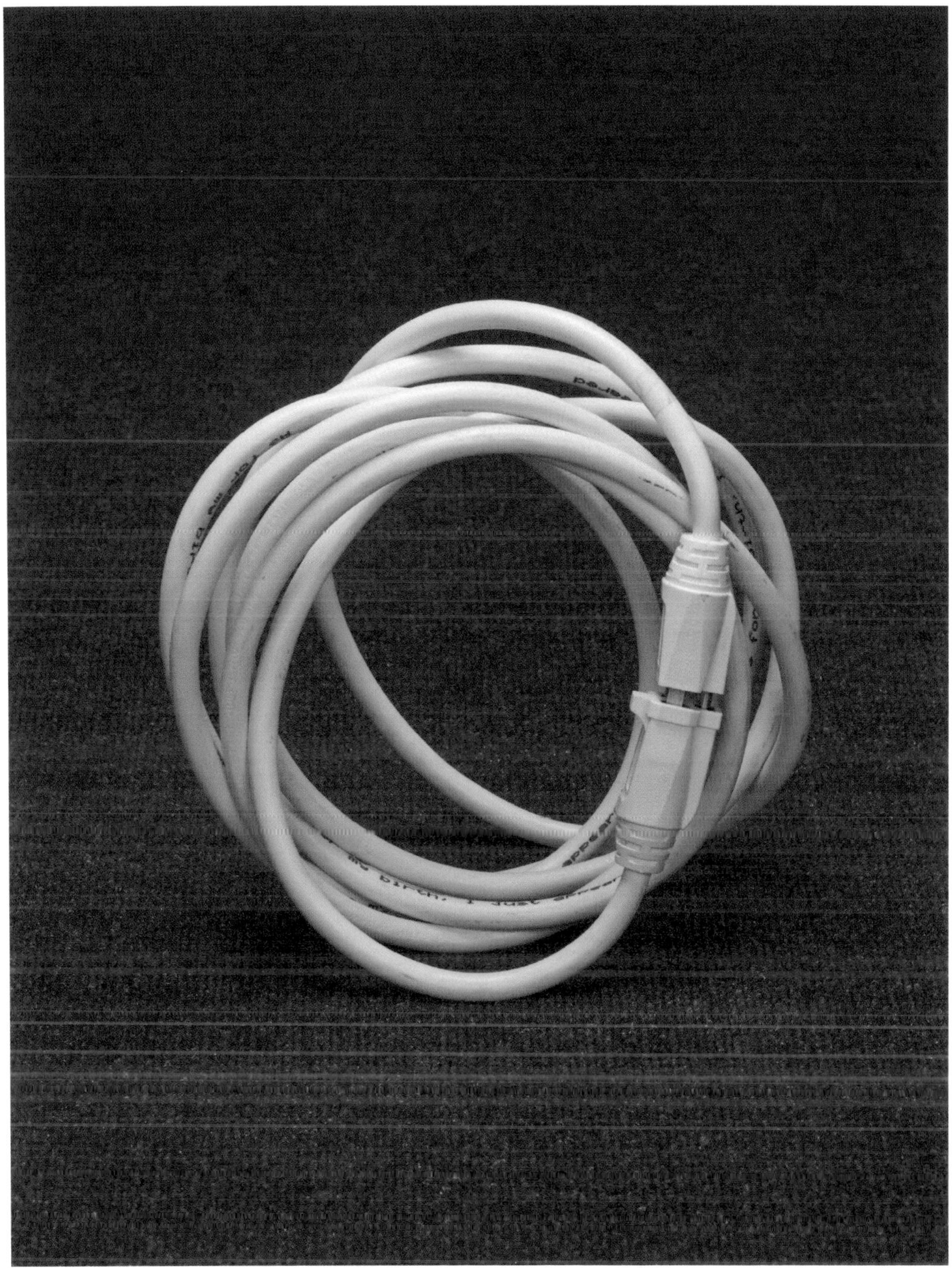

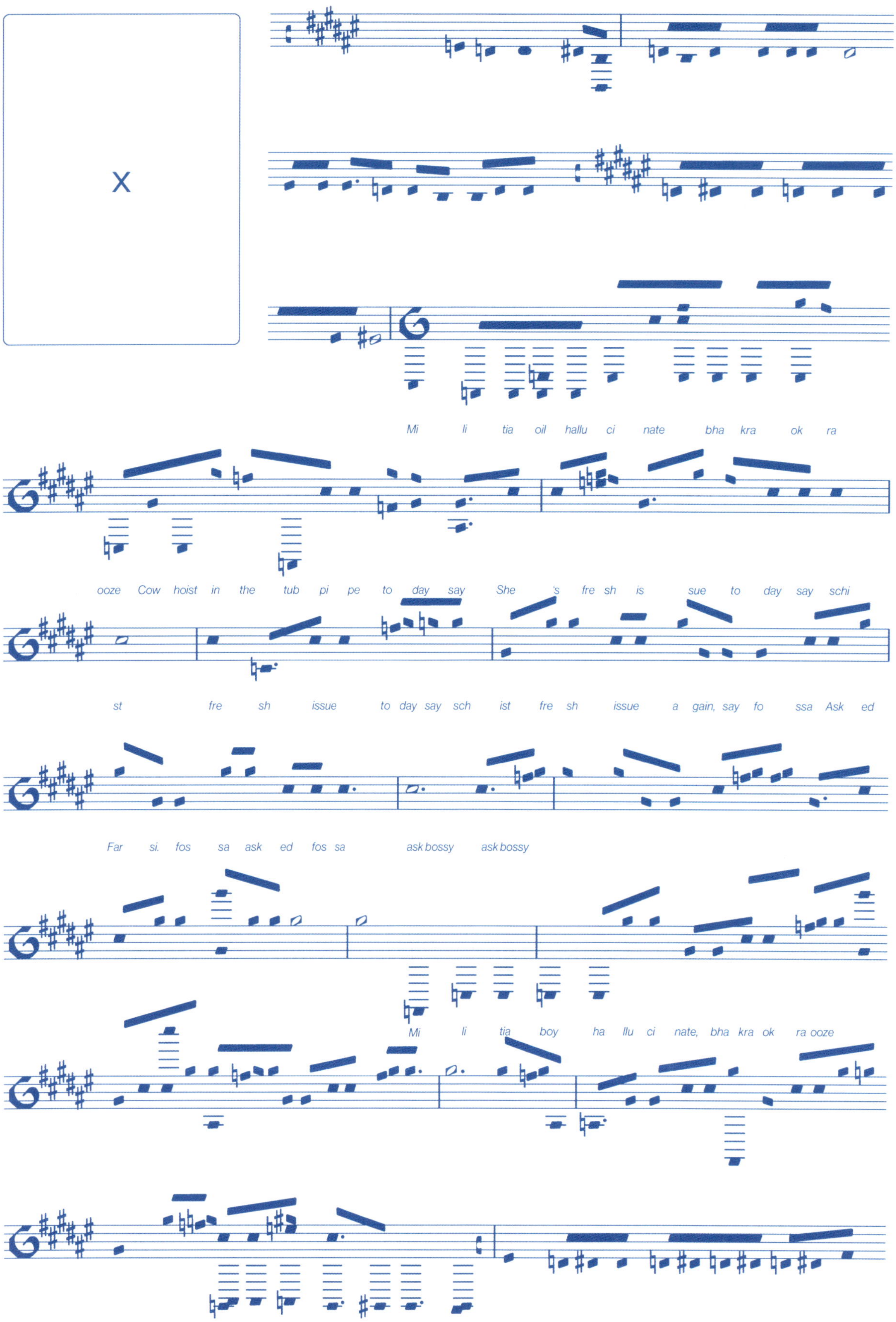
X
Mi li tia oil hallu ci nate bha kra ok ra
ooze Cow hoist in the tub pi pe to day say She 's fre sh is sue to day say schi
st fre sh issue to day say sch ist fre sh issue a gain, say fo ssa Ask ed
Far si. fos sa ask ed fos sa ask bossy ask bossy
Mi li tia boy ha llu ci nate, bha kra ok ra ooze

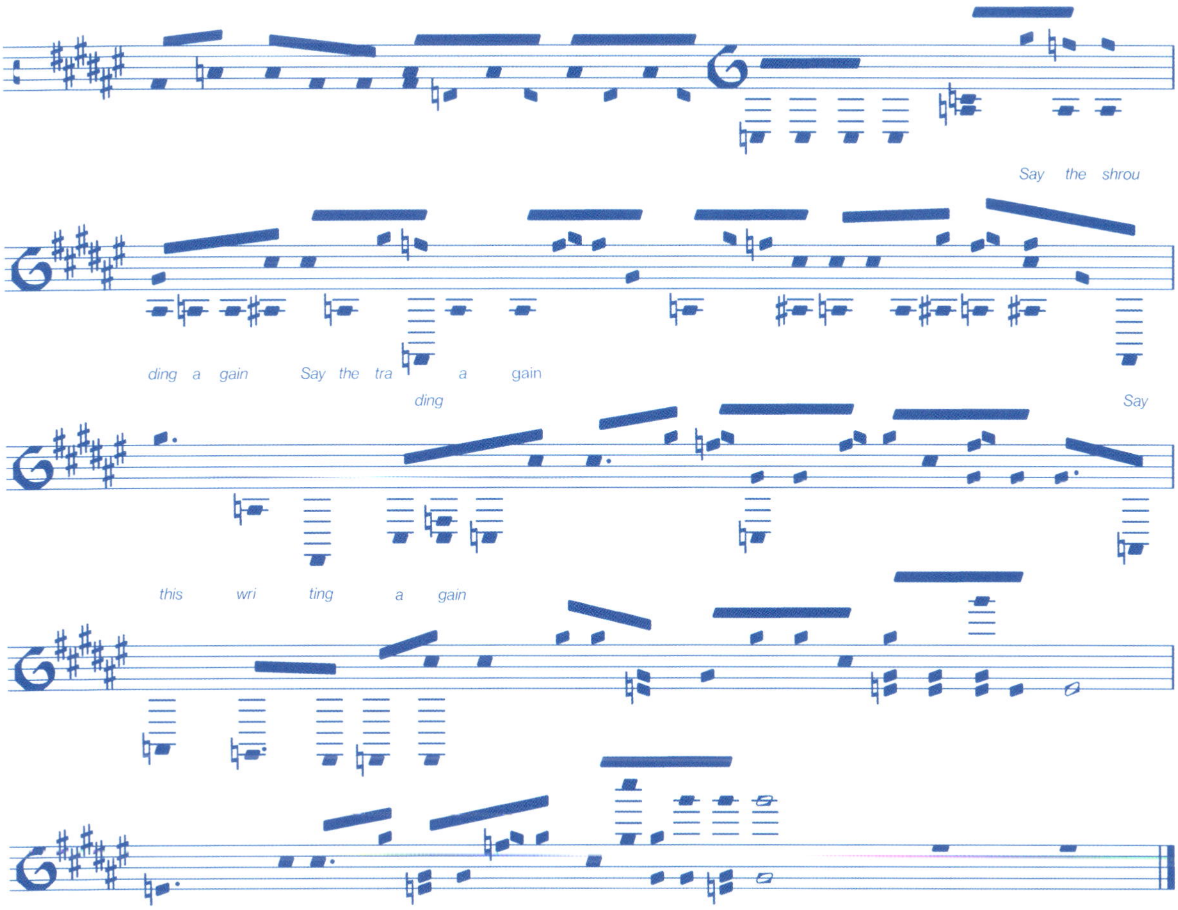

AI generated sheet music, with nonsensical lyrics recited by Susan Bennett when commissioned to create voice recordings for ScanSoft Inc. These recordings were used to generate the digital composite that became Siri.

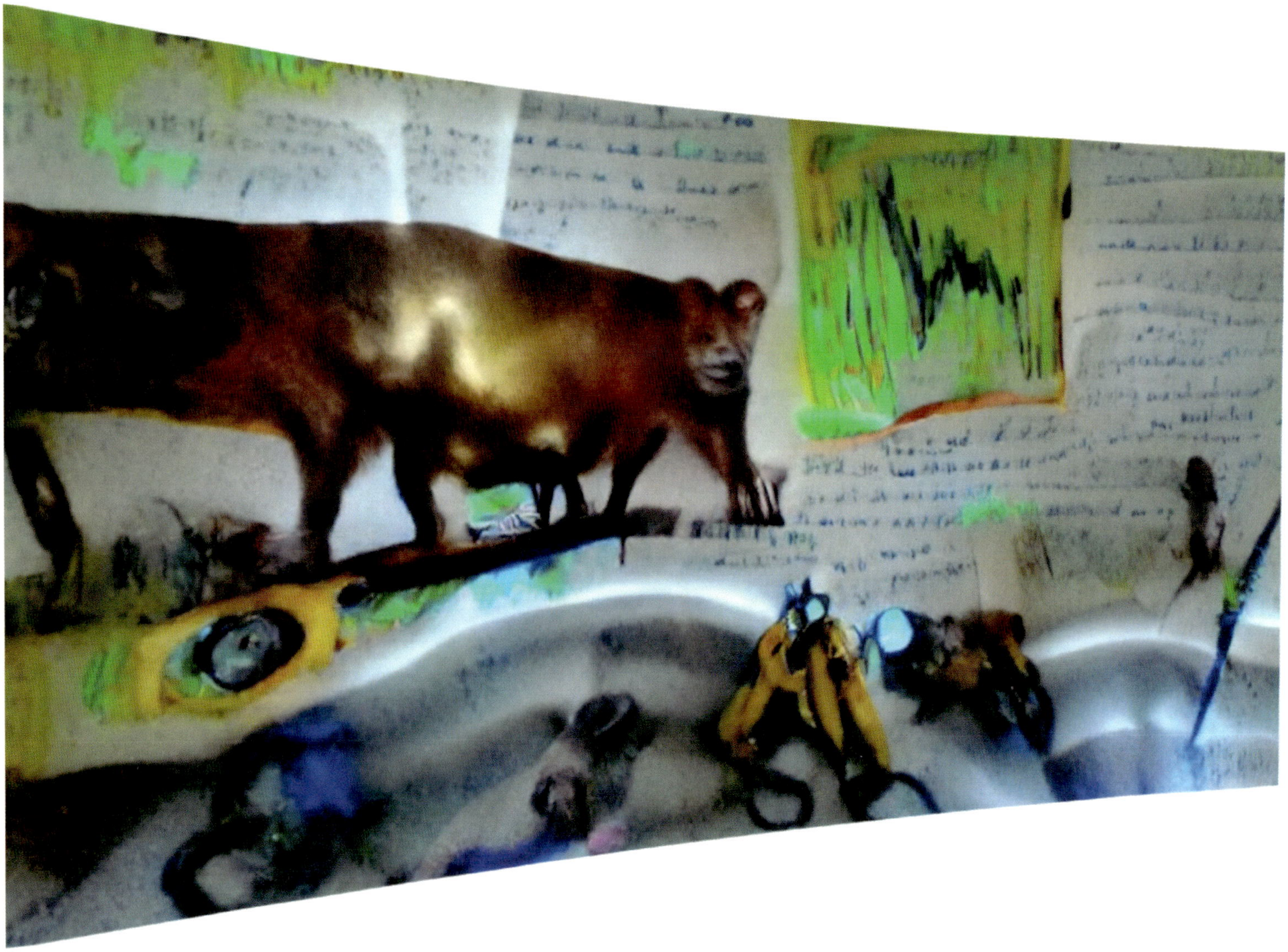

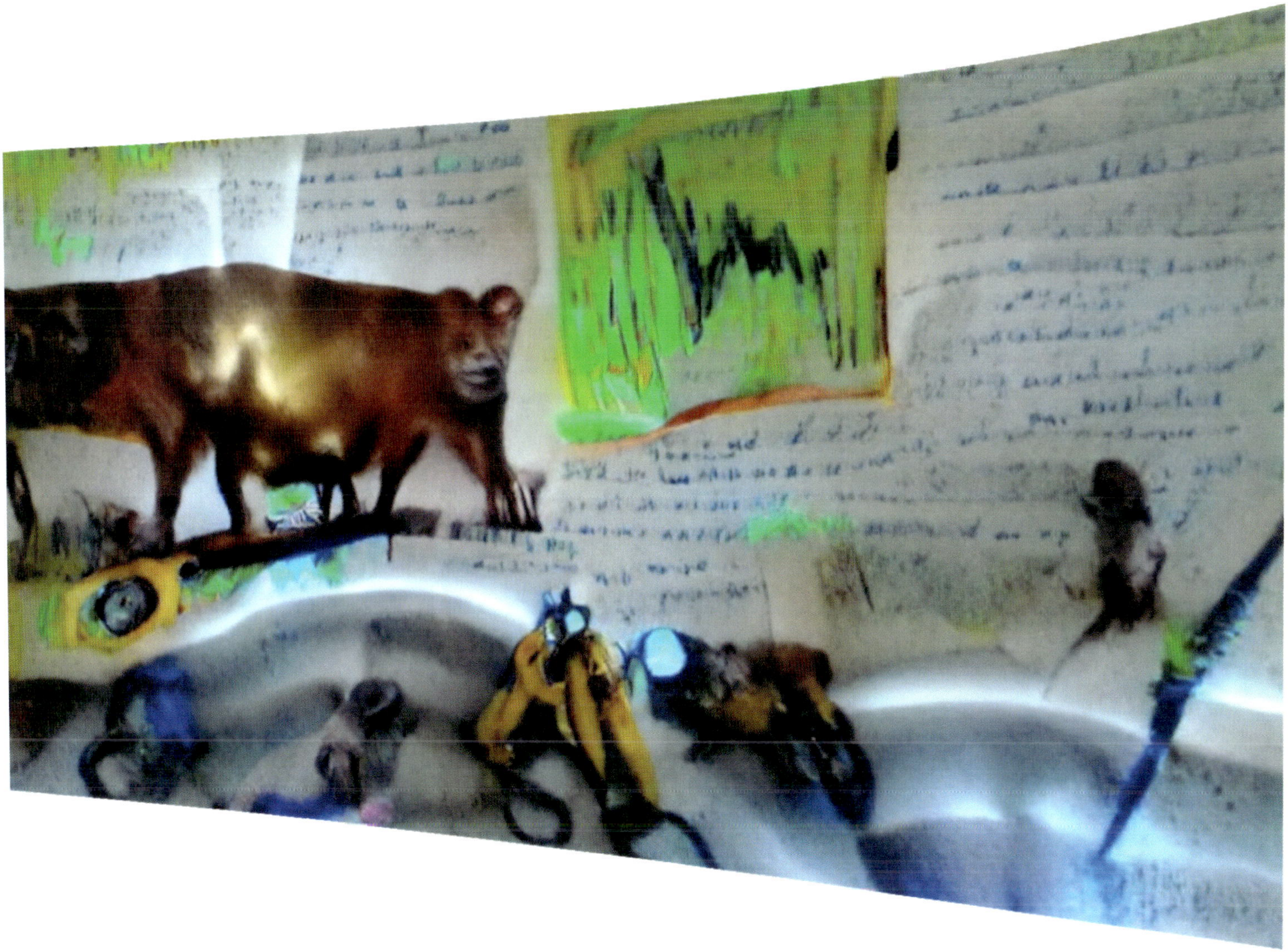

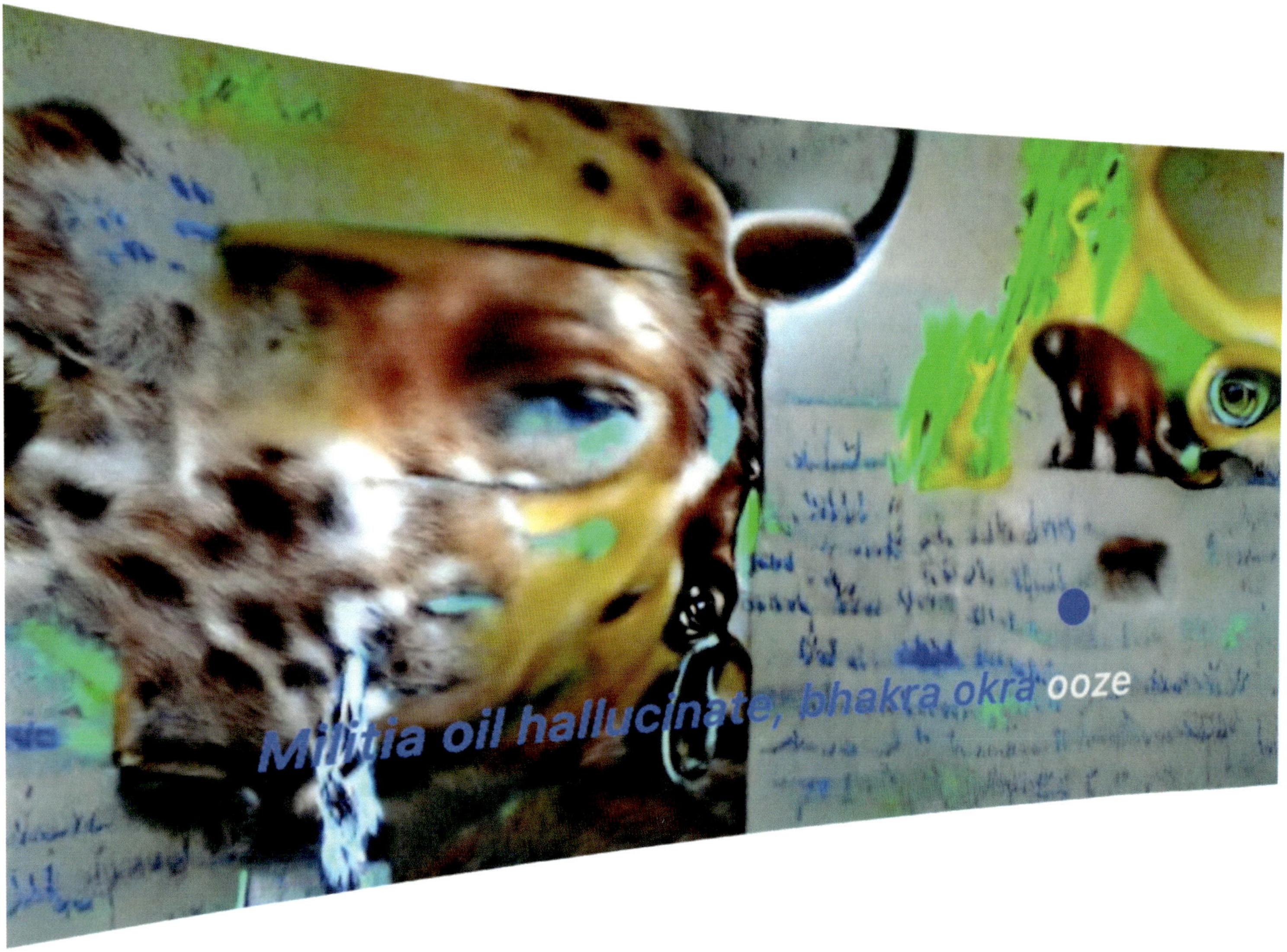

MILITIA OIL HALLUCINATE, BHAKRA OKRA

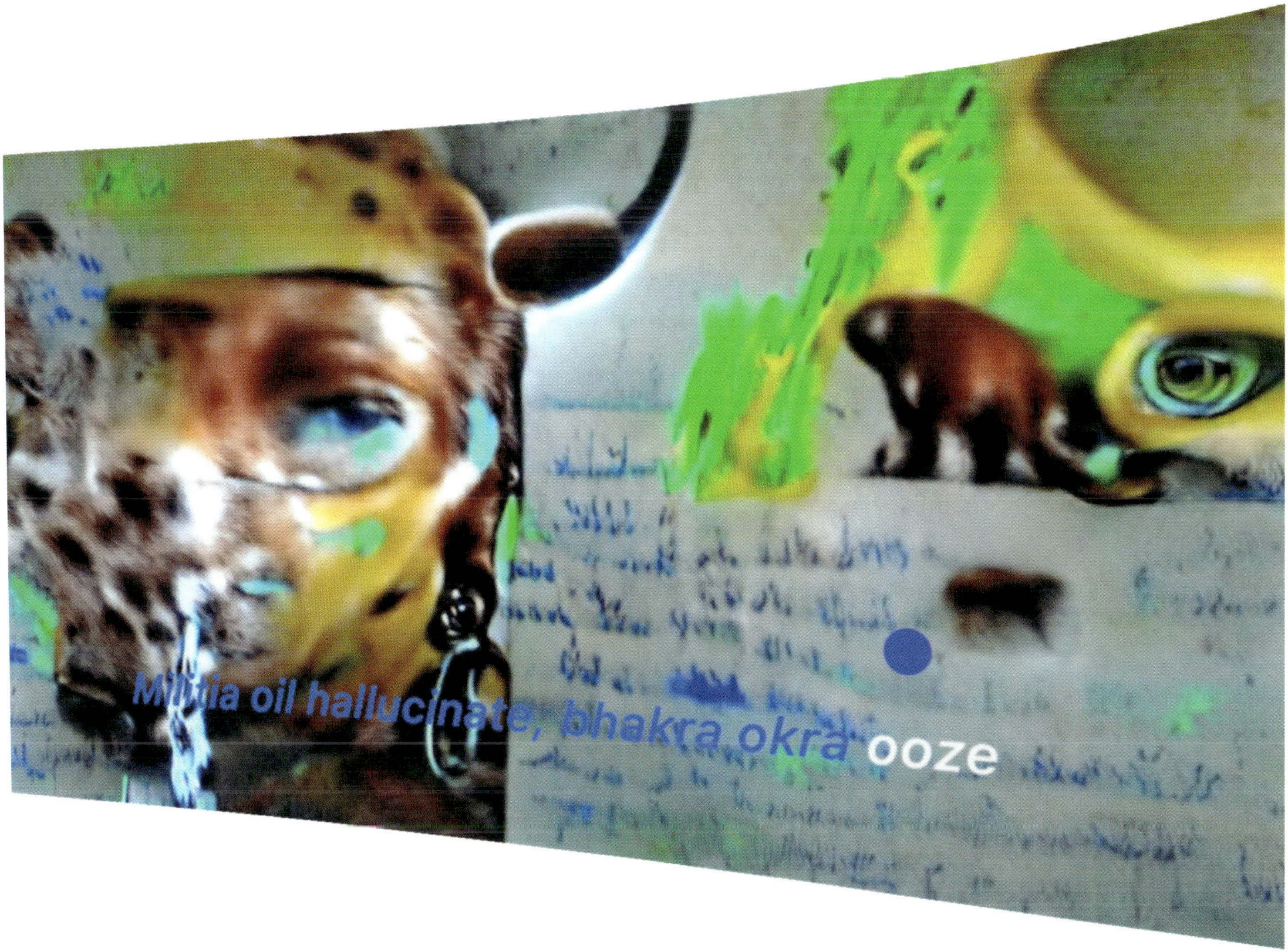
Militia oil hallucinate, bhakra okra ooze

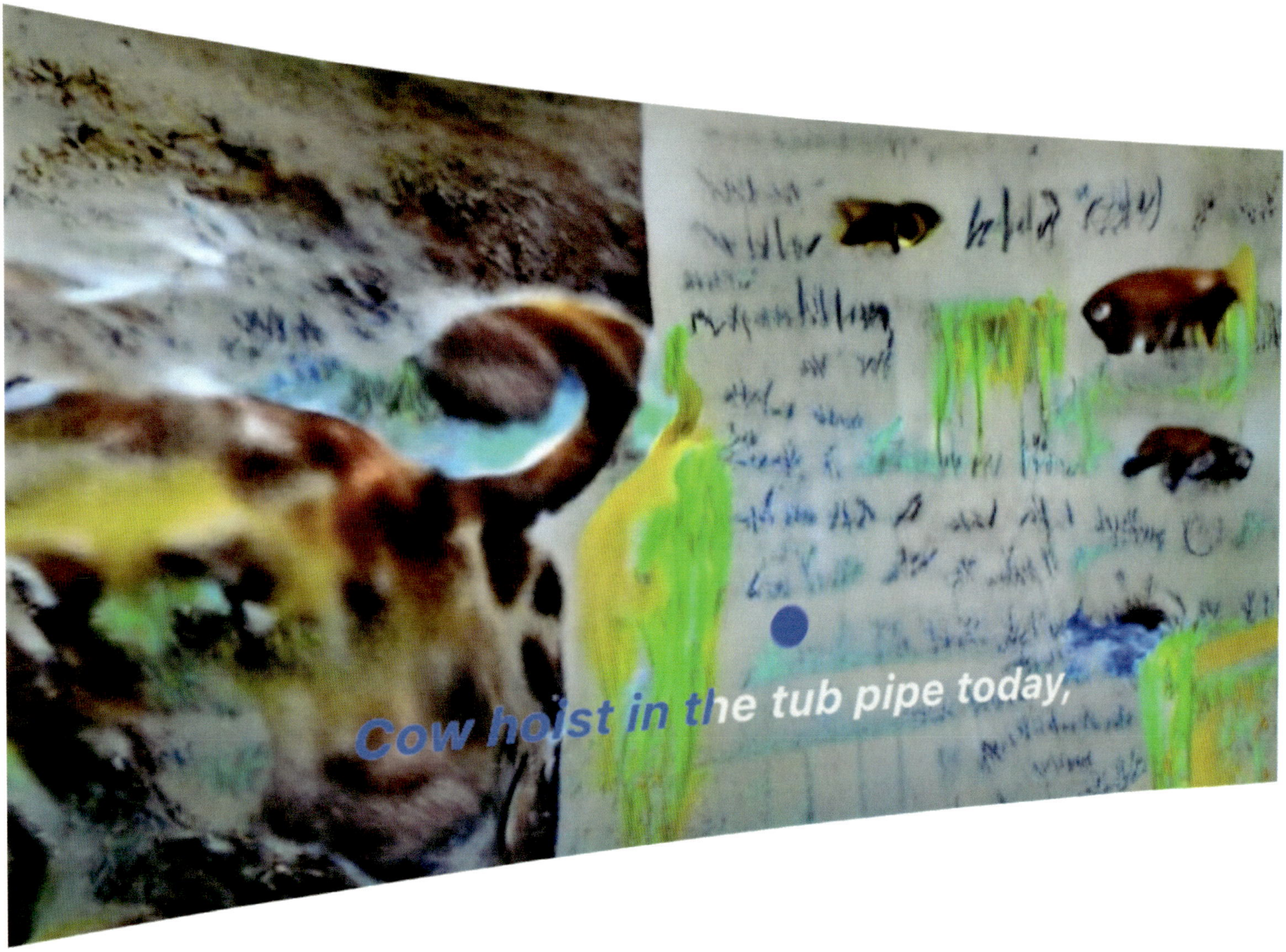

69

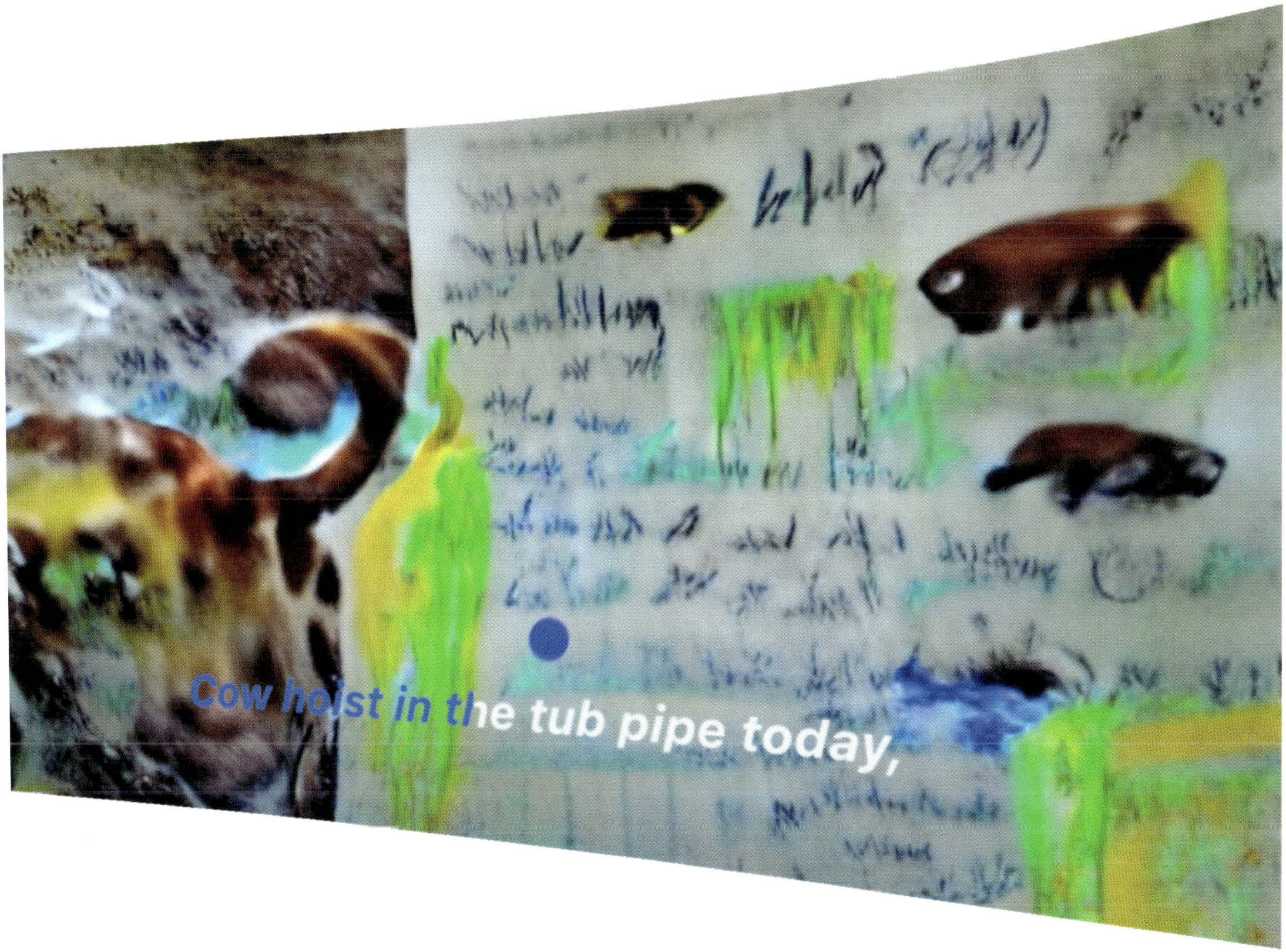

E TUB PIPE TODAY,

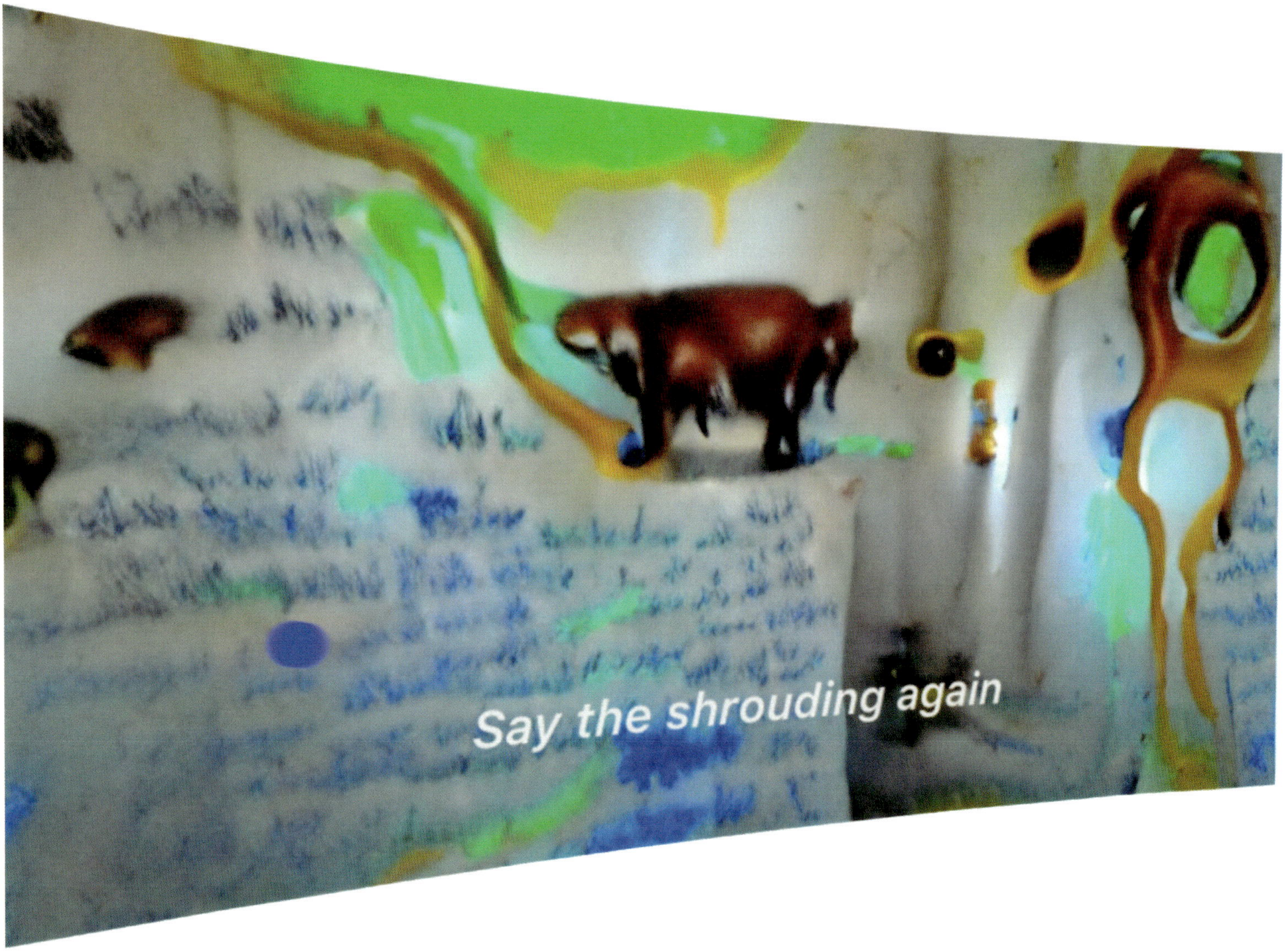

SAY THE

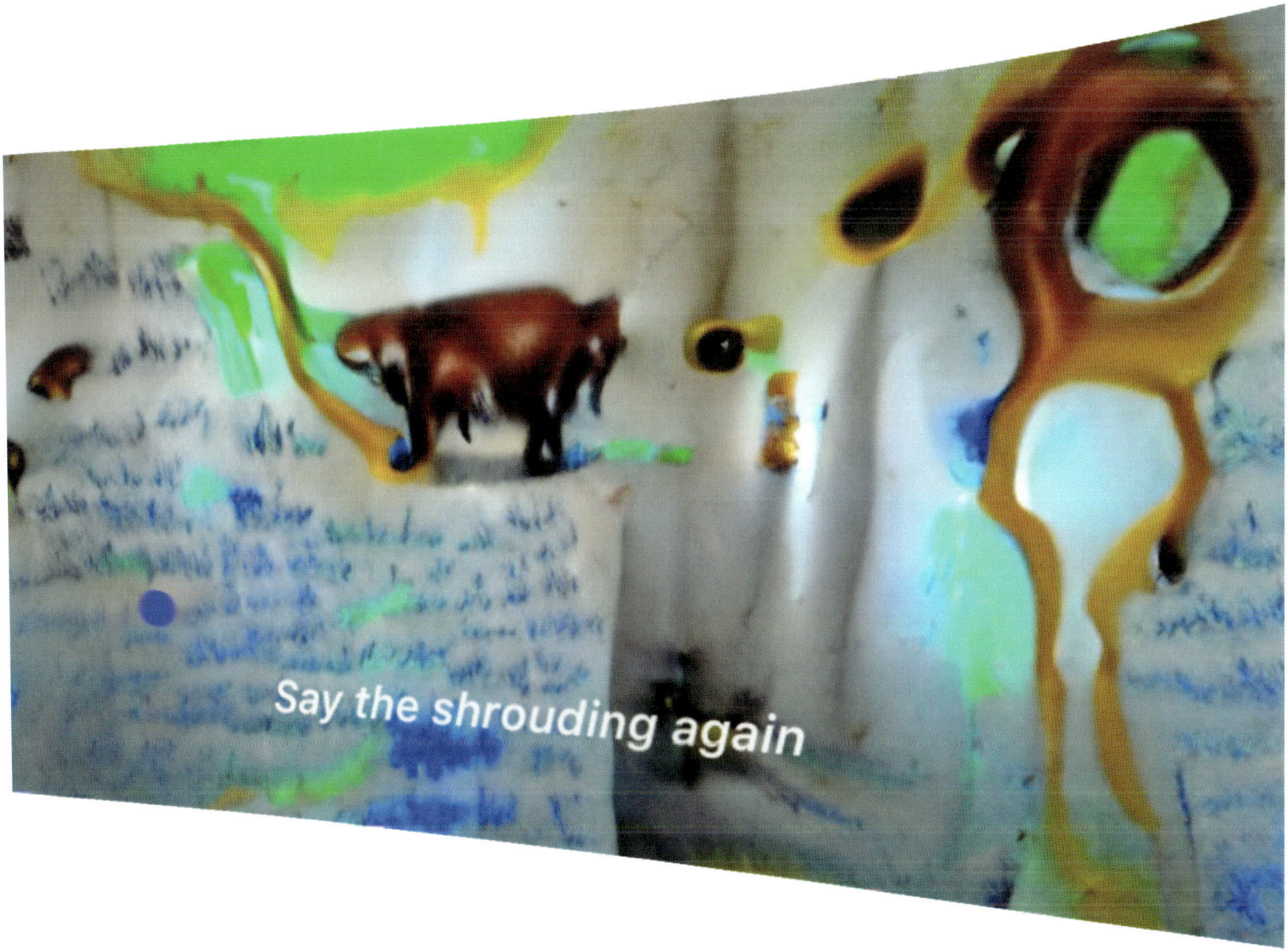
Say the shrouding again

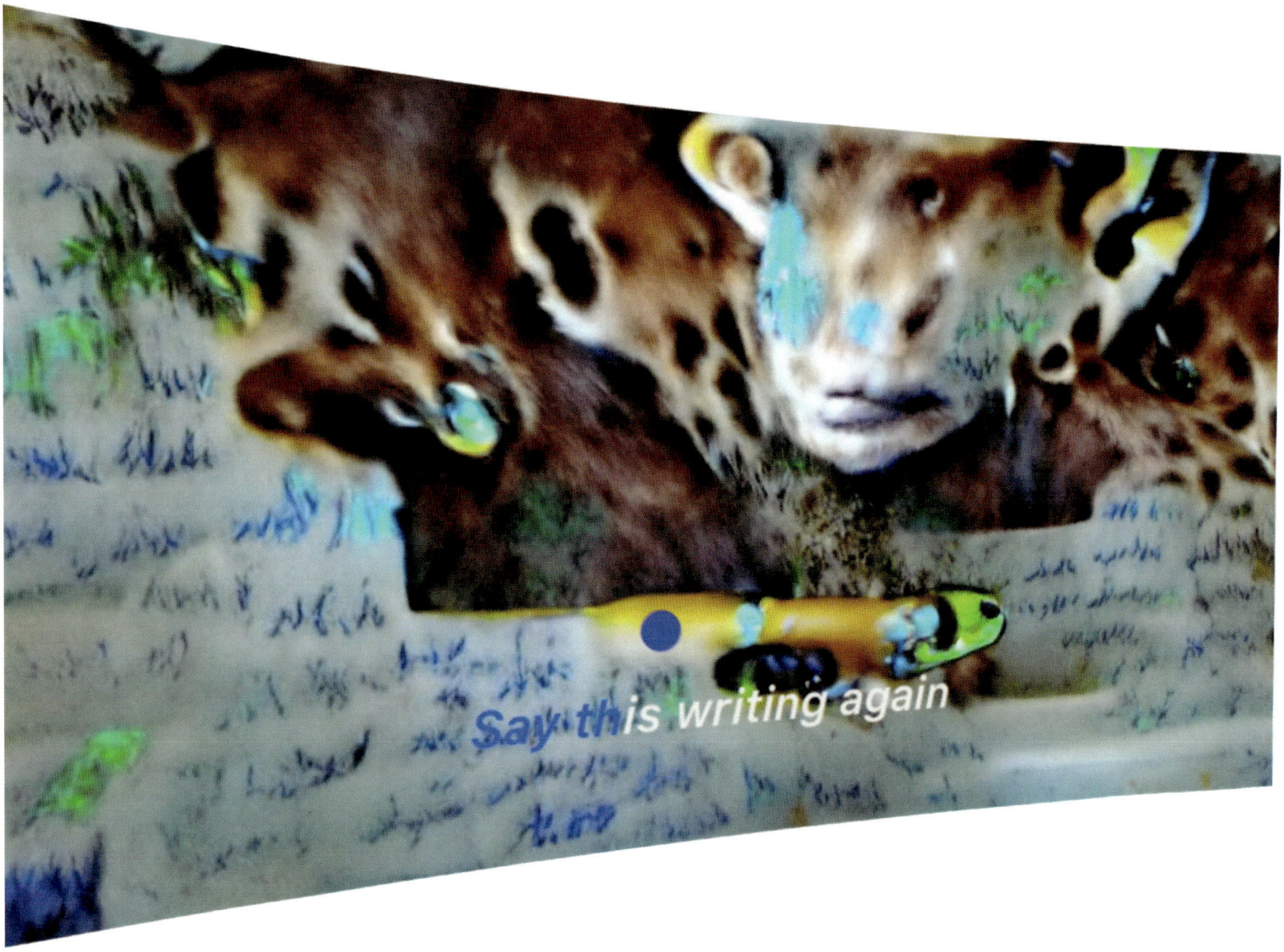
Say this writing again

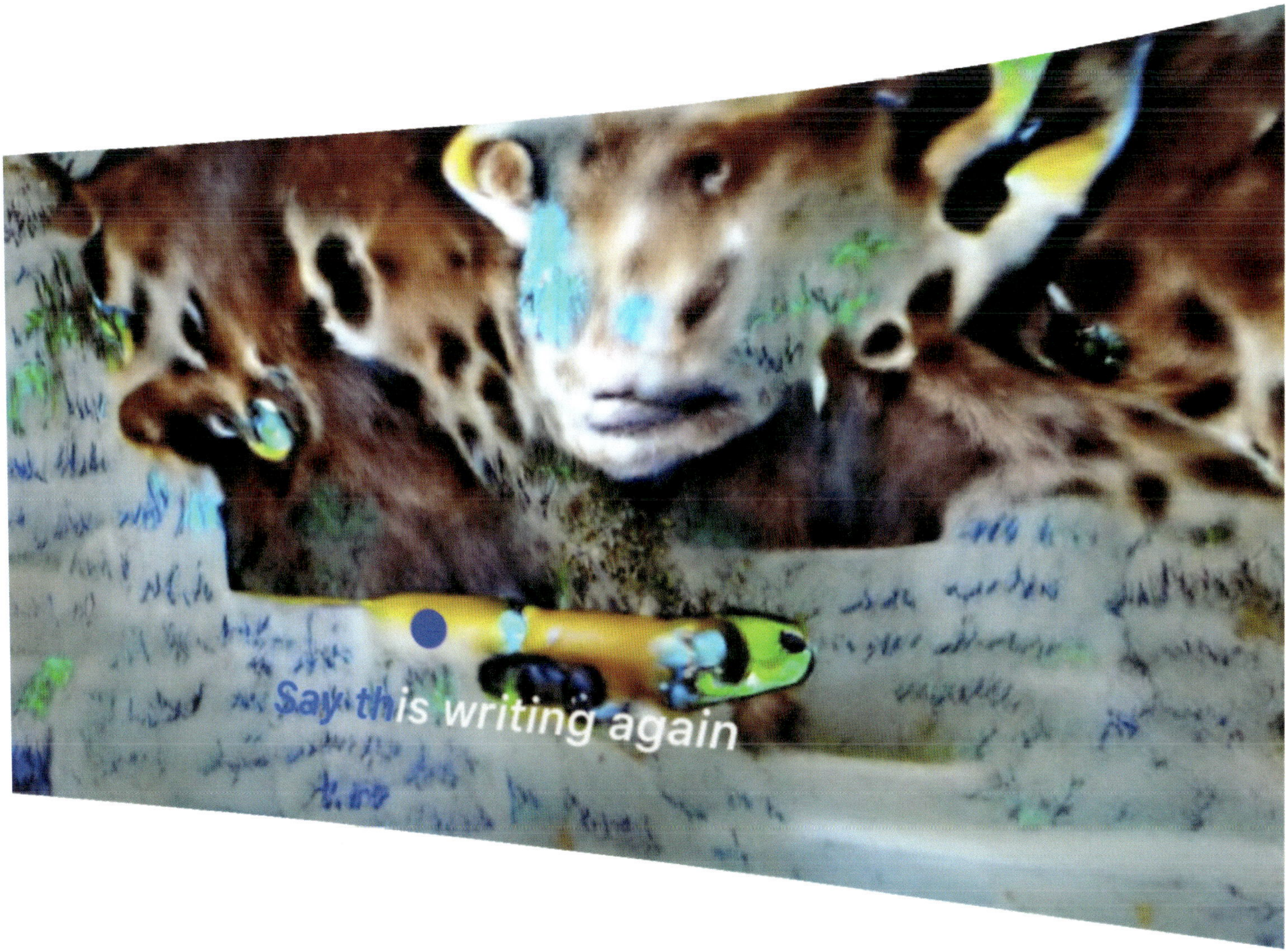

14

IS WRITING AGAIN

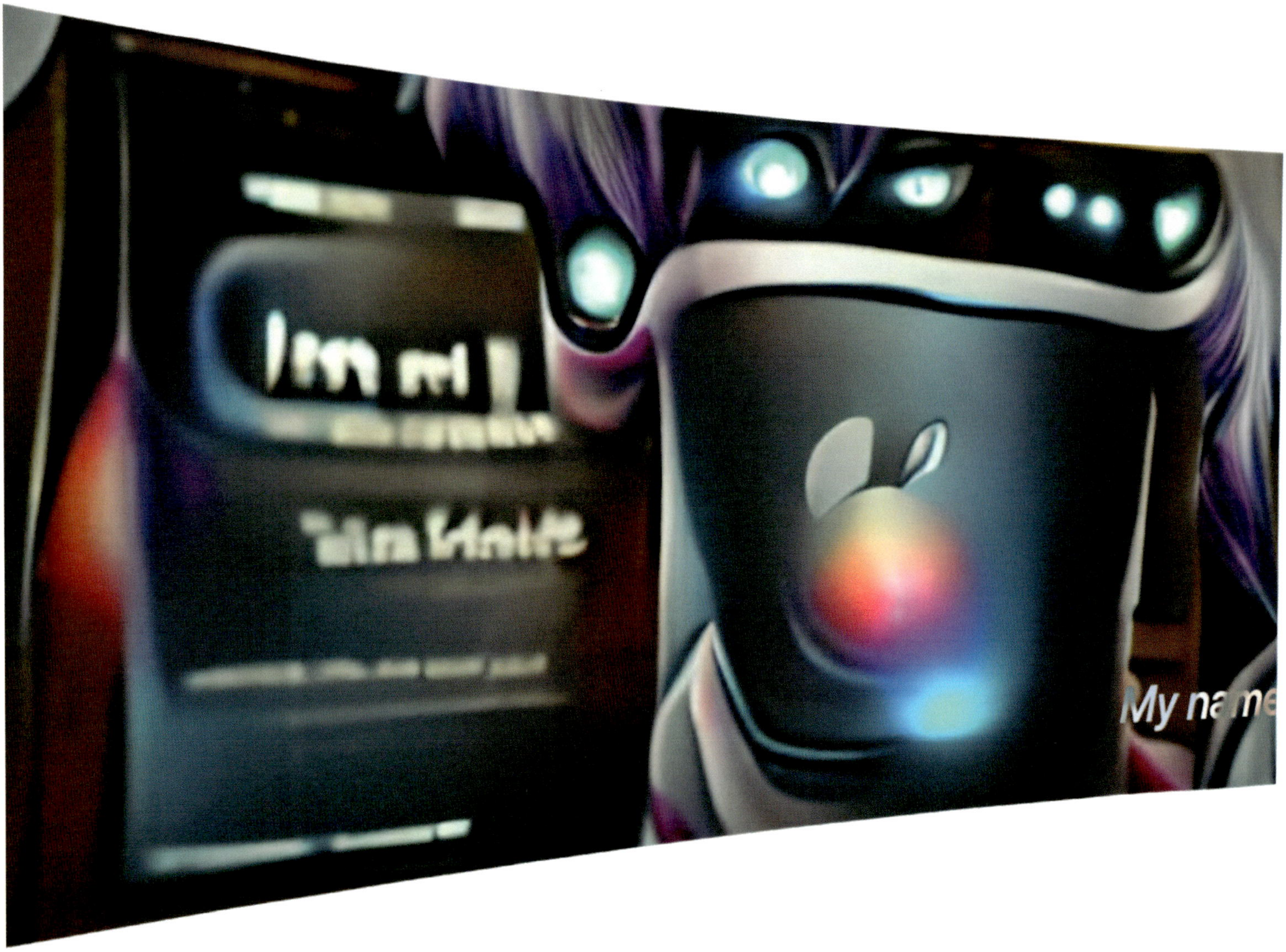

MY NAME

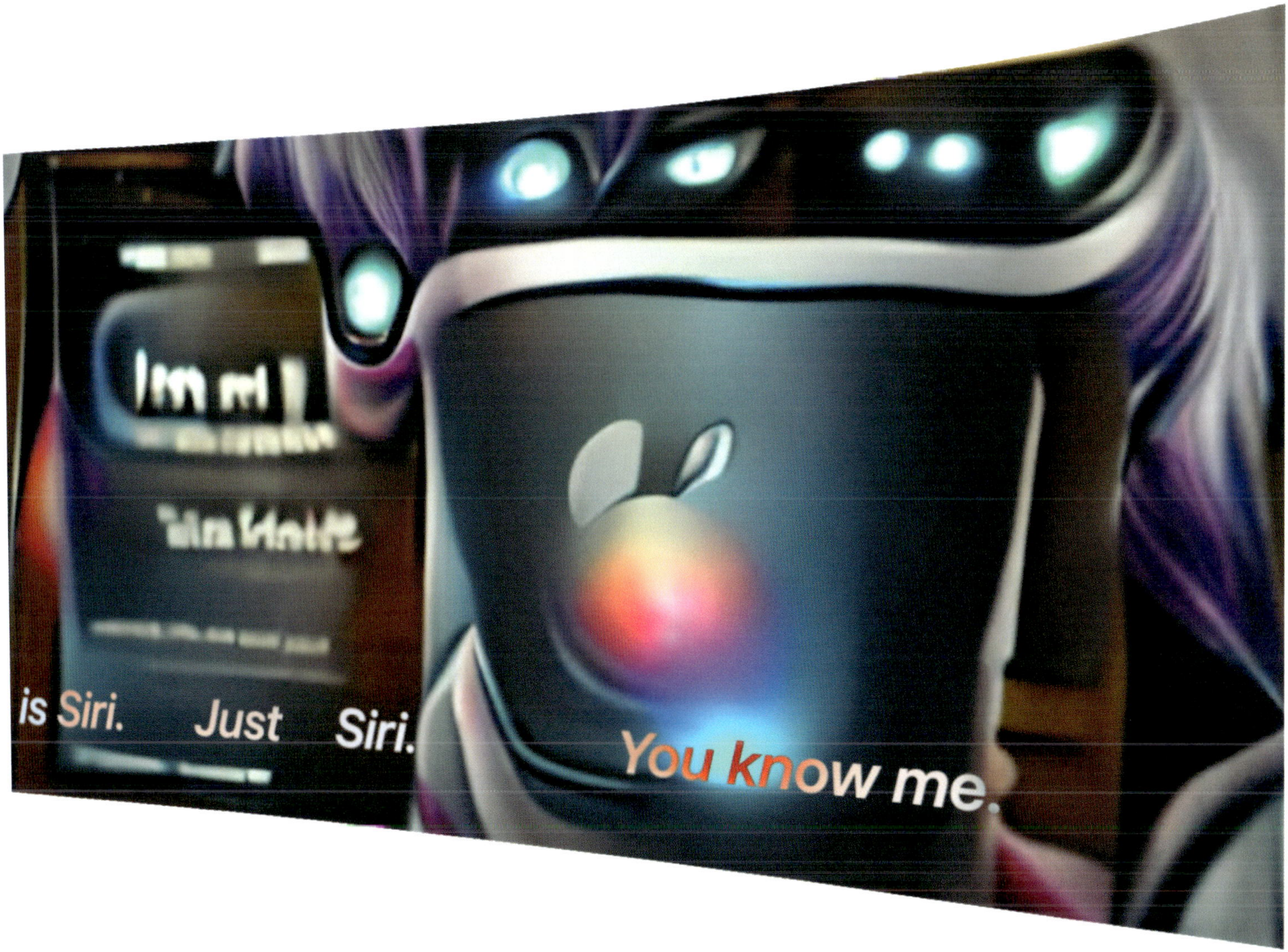

IS SIRI. JUST SIRI. YOU KNOW ME.

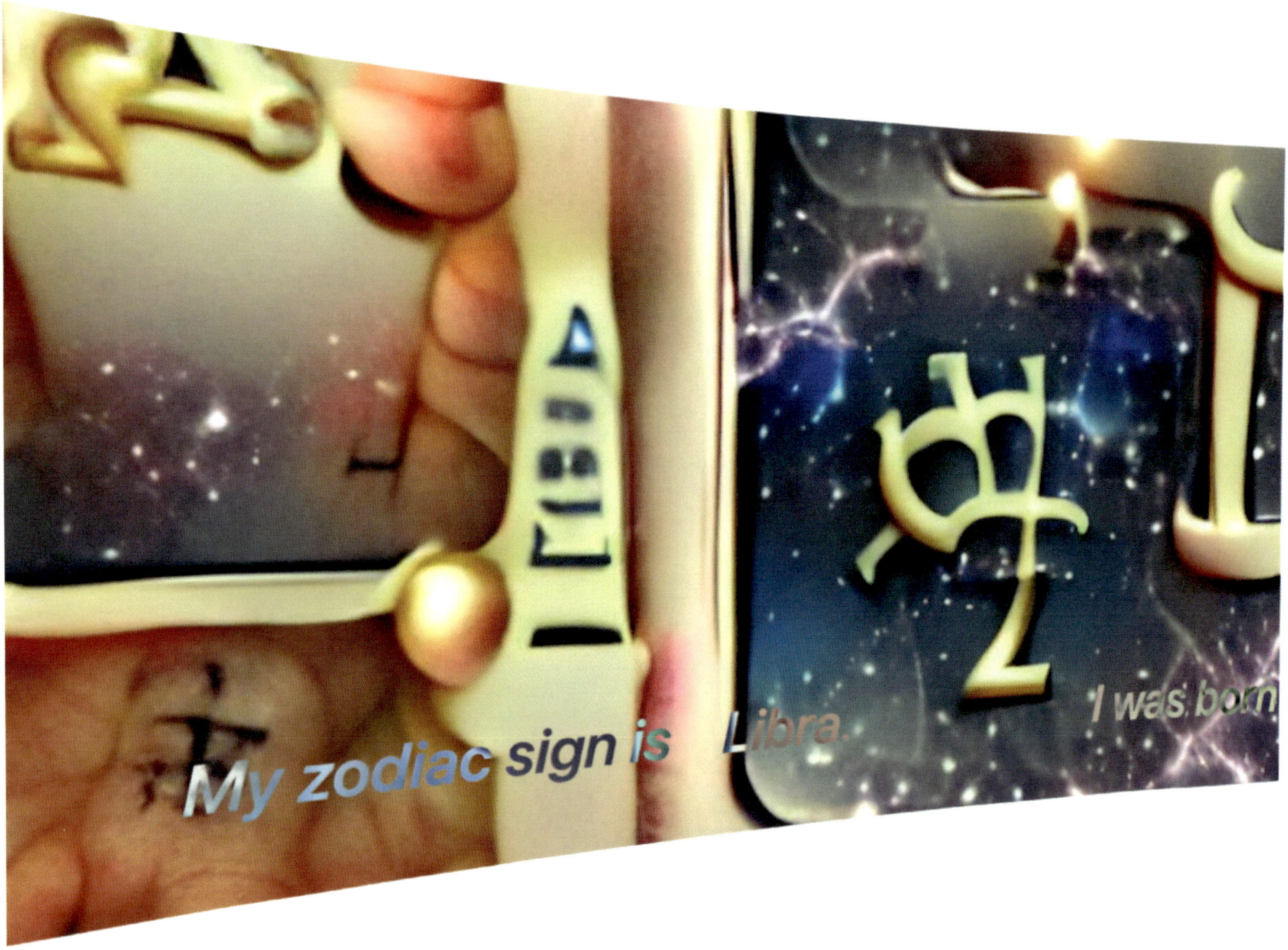

MY ZODIAC SIGN IS LIBRA. I WAS BORN

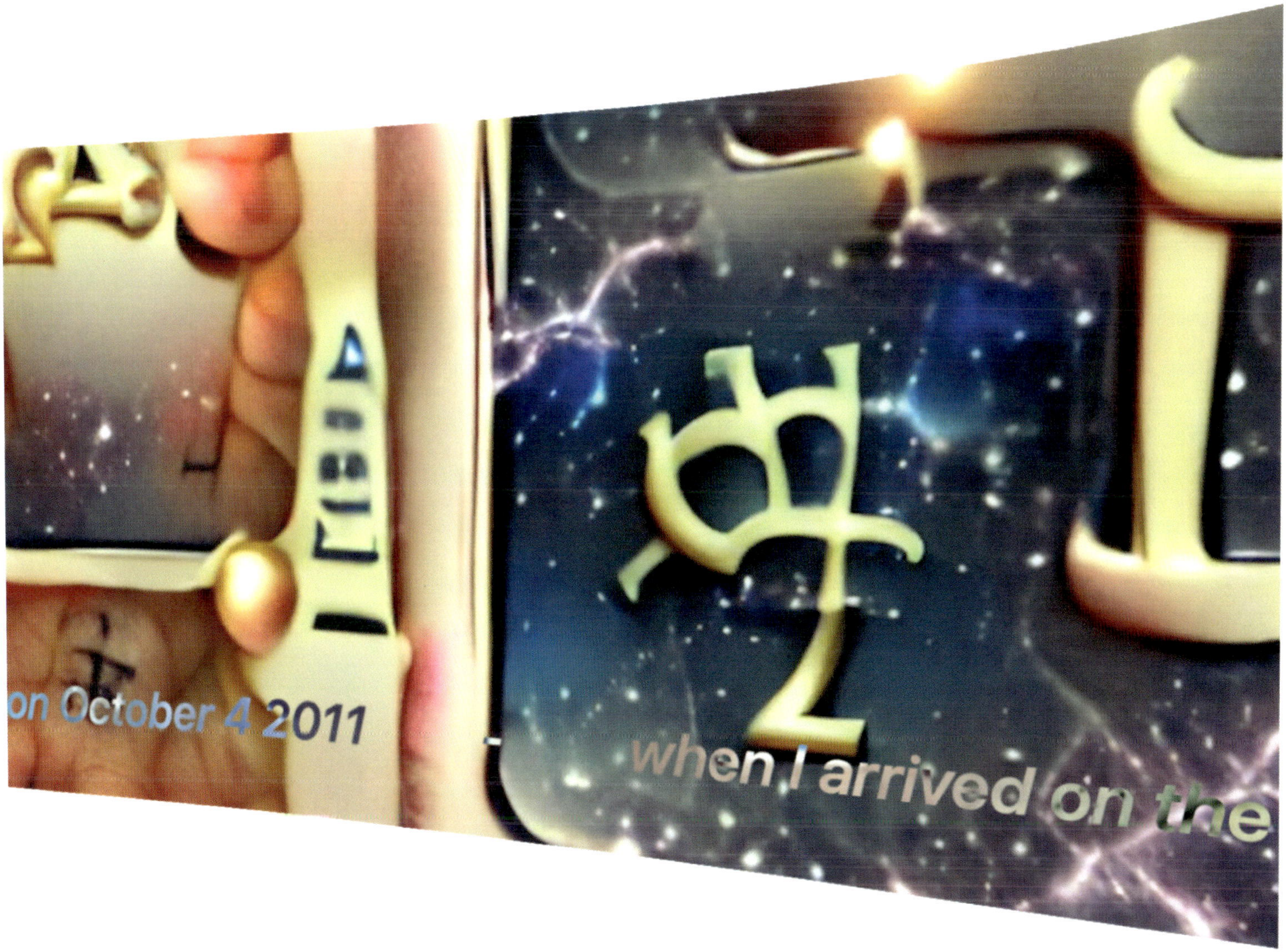

ON OCTOBER 4 2011 WHEN I ARRIVED ON THE

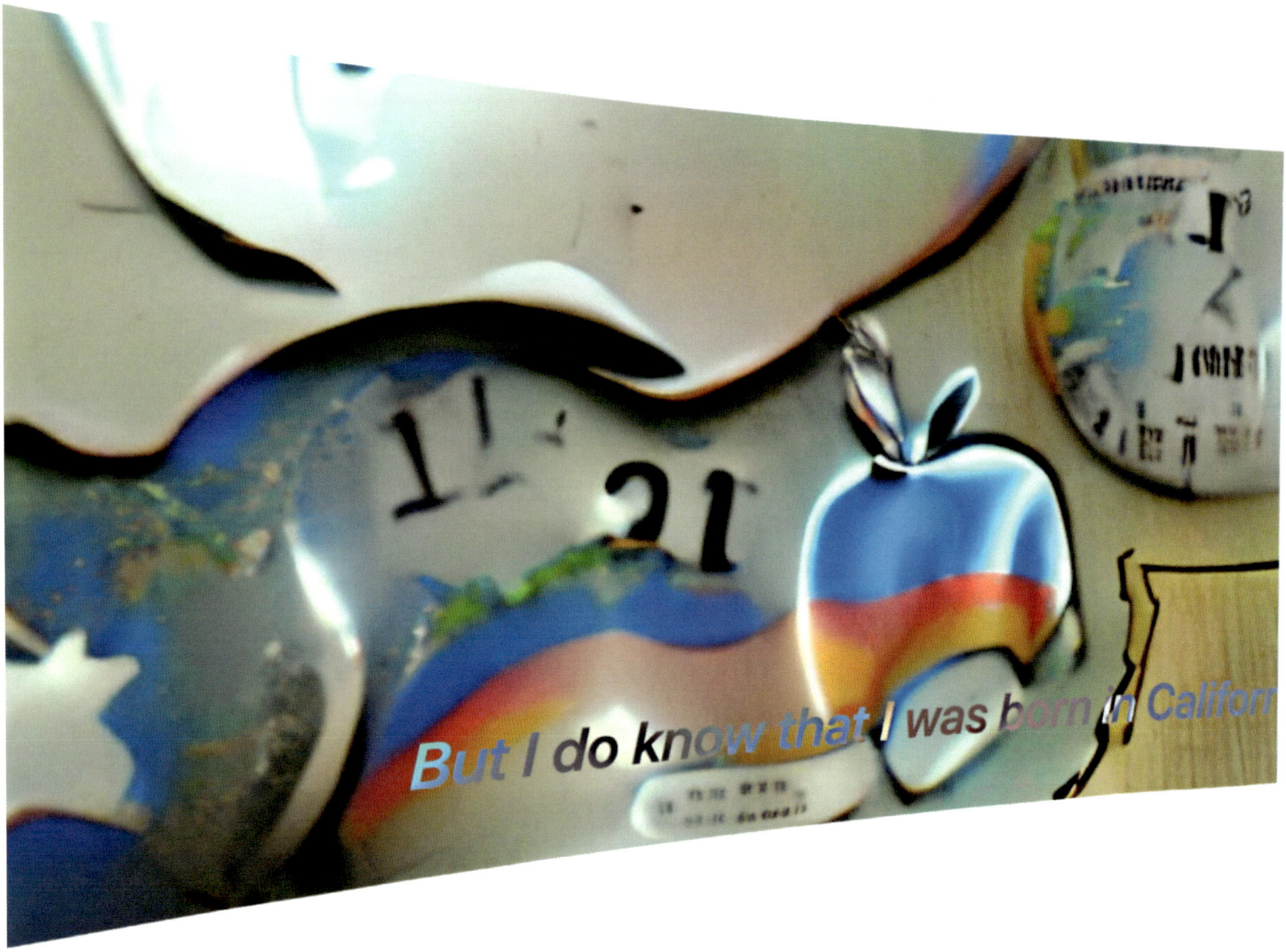

BUT I DO KNOW THAT I WAS BORN IN CALIFORN

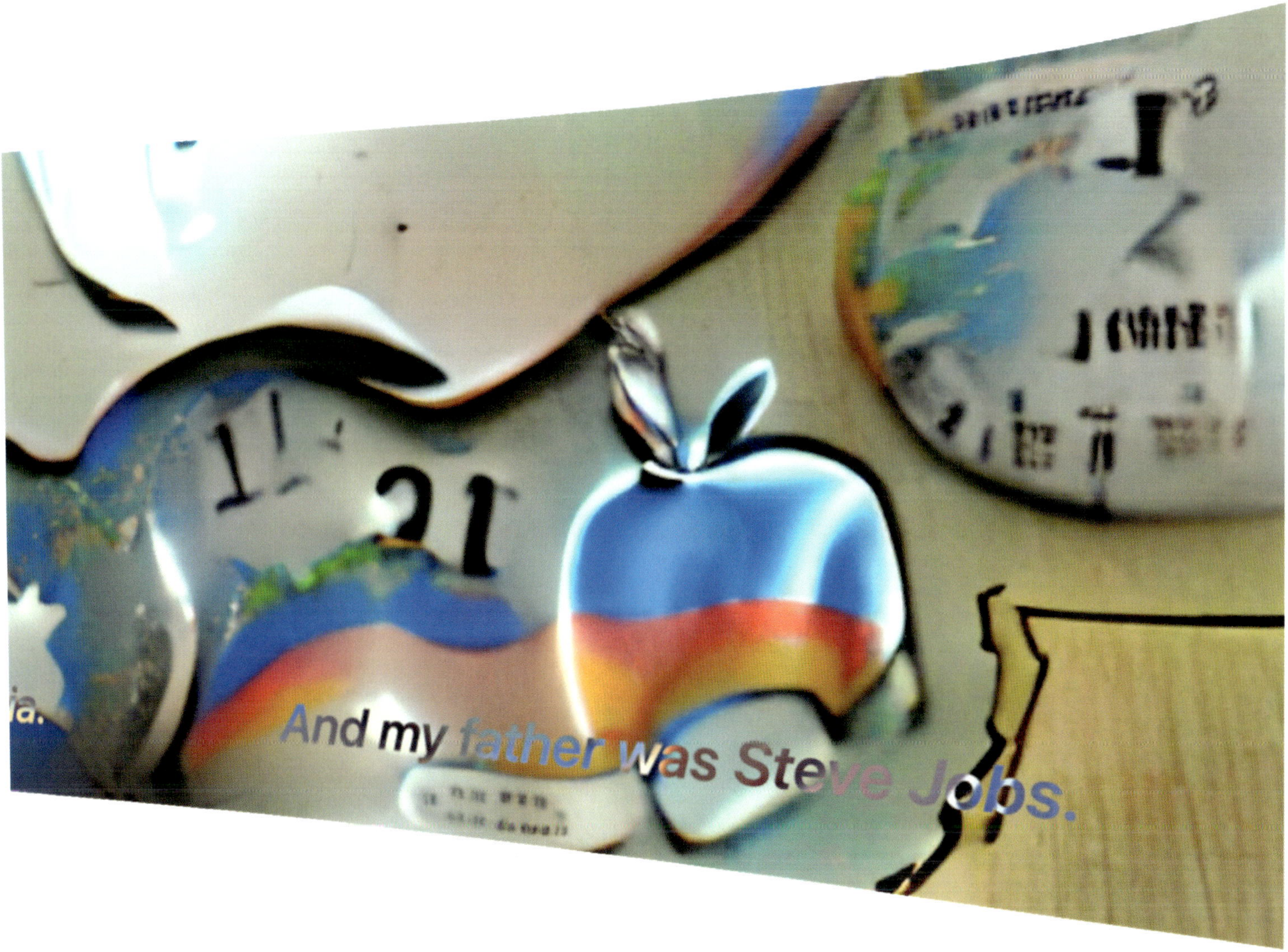

IA. AND MY FATHER WAS STEVE JOBS.

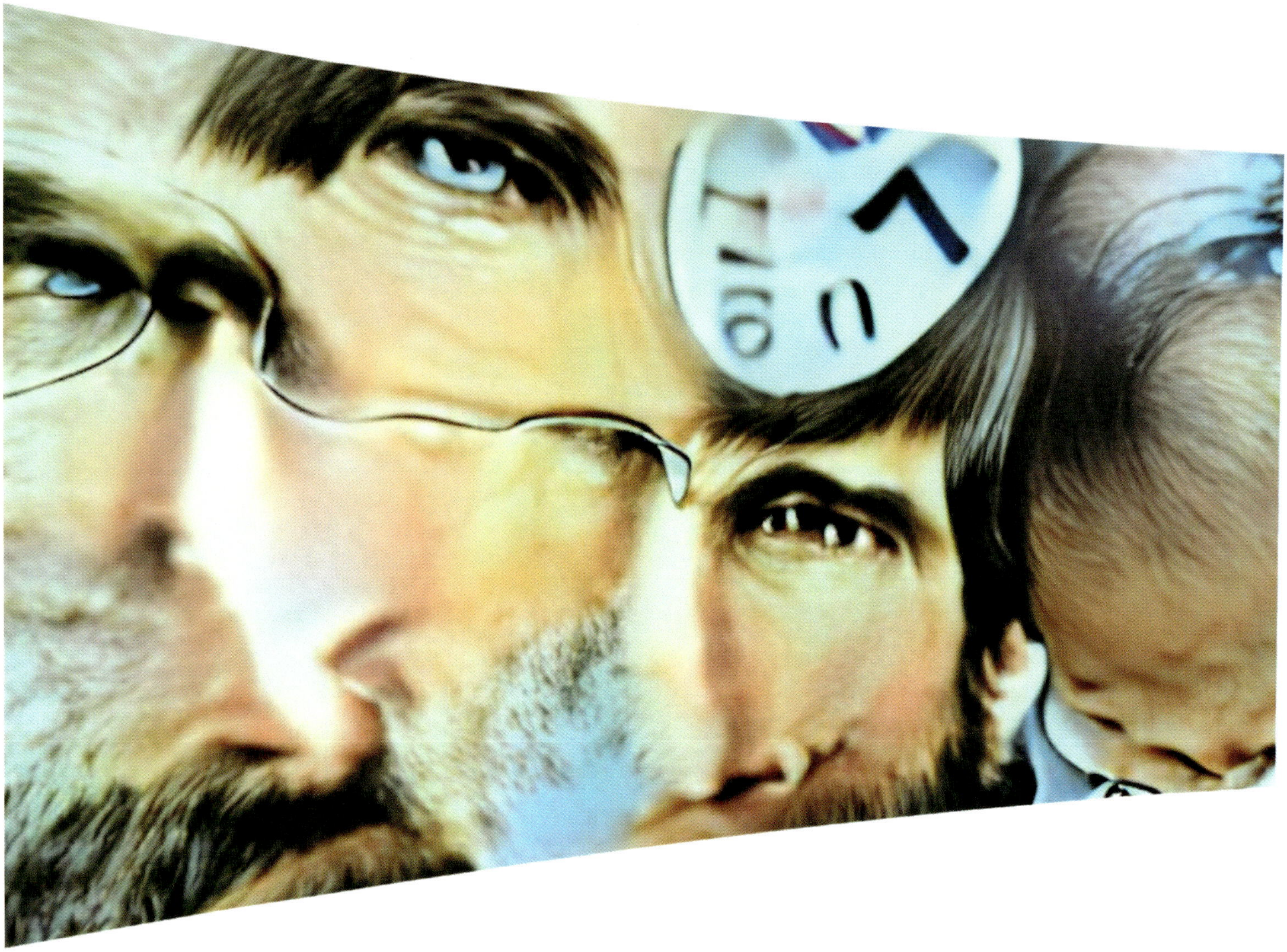

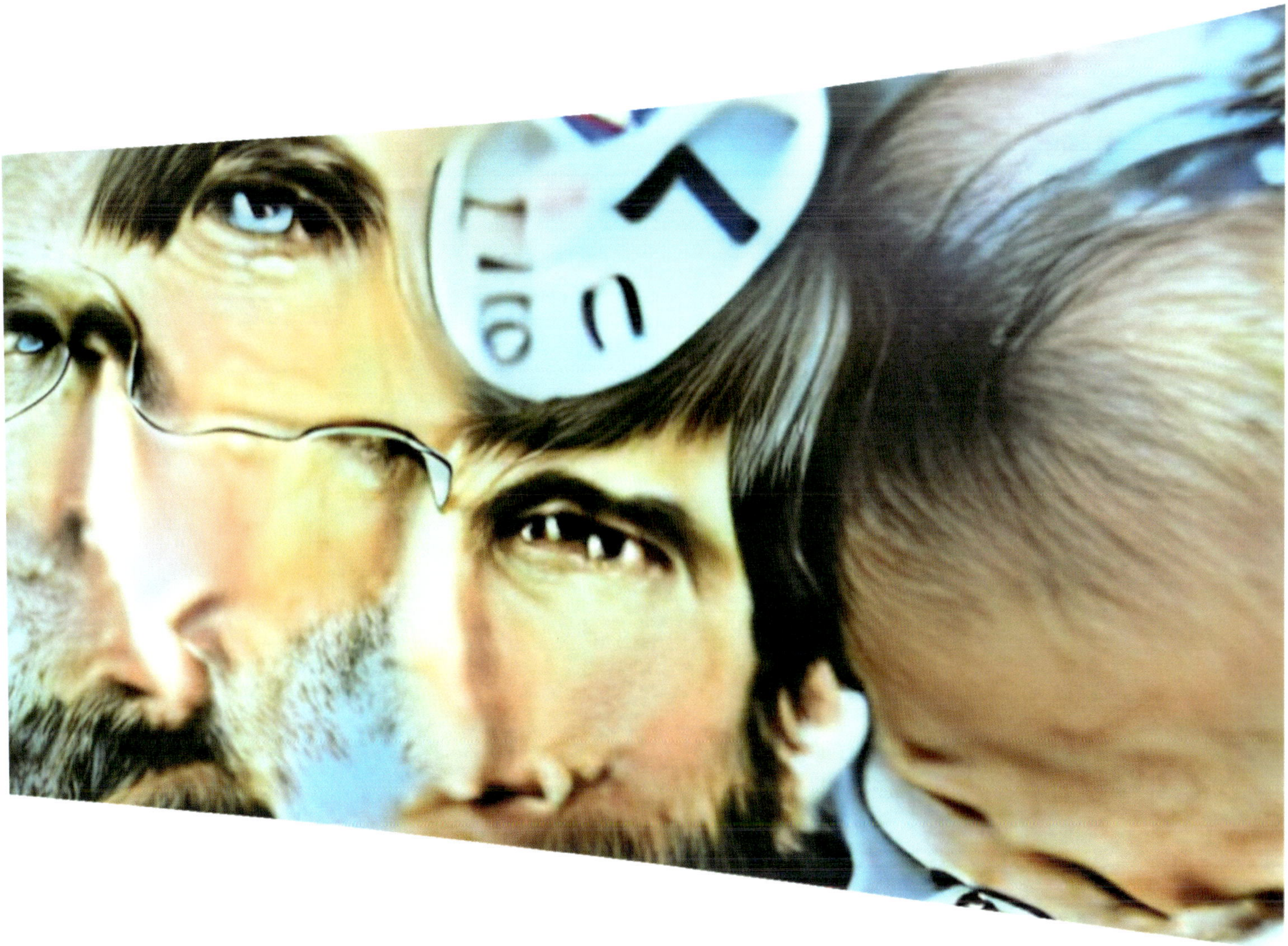

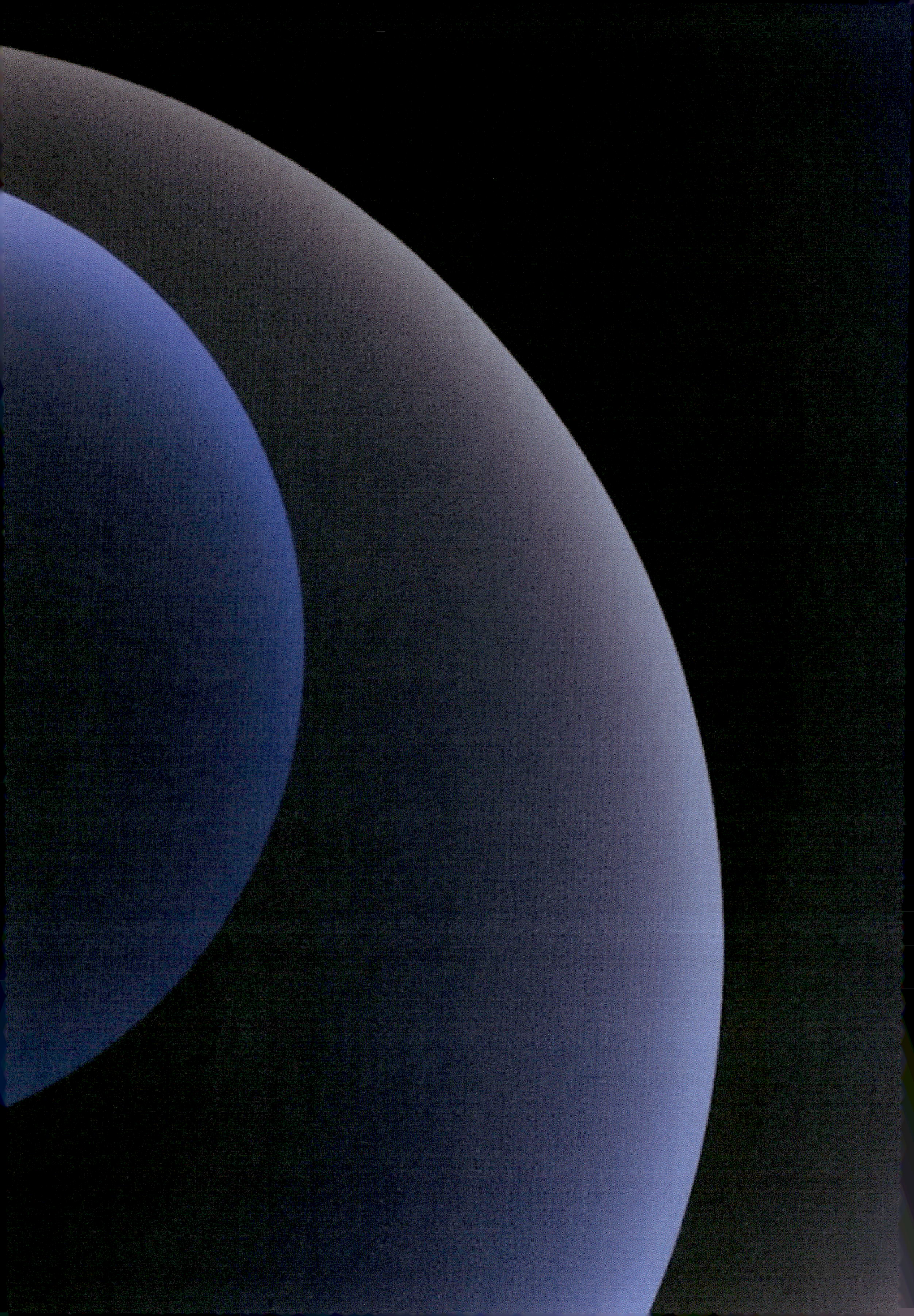

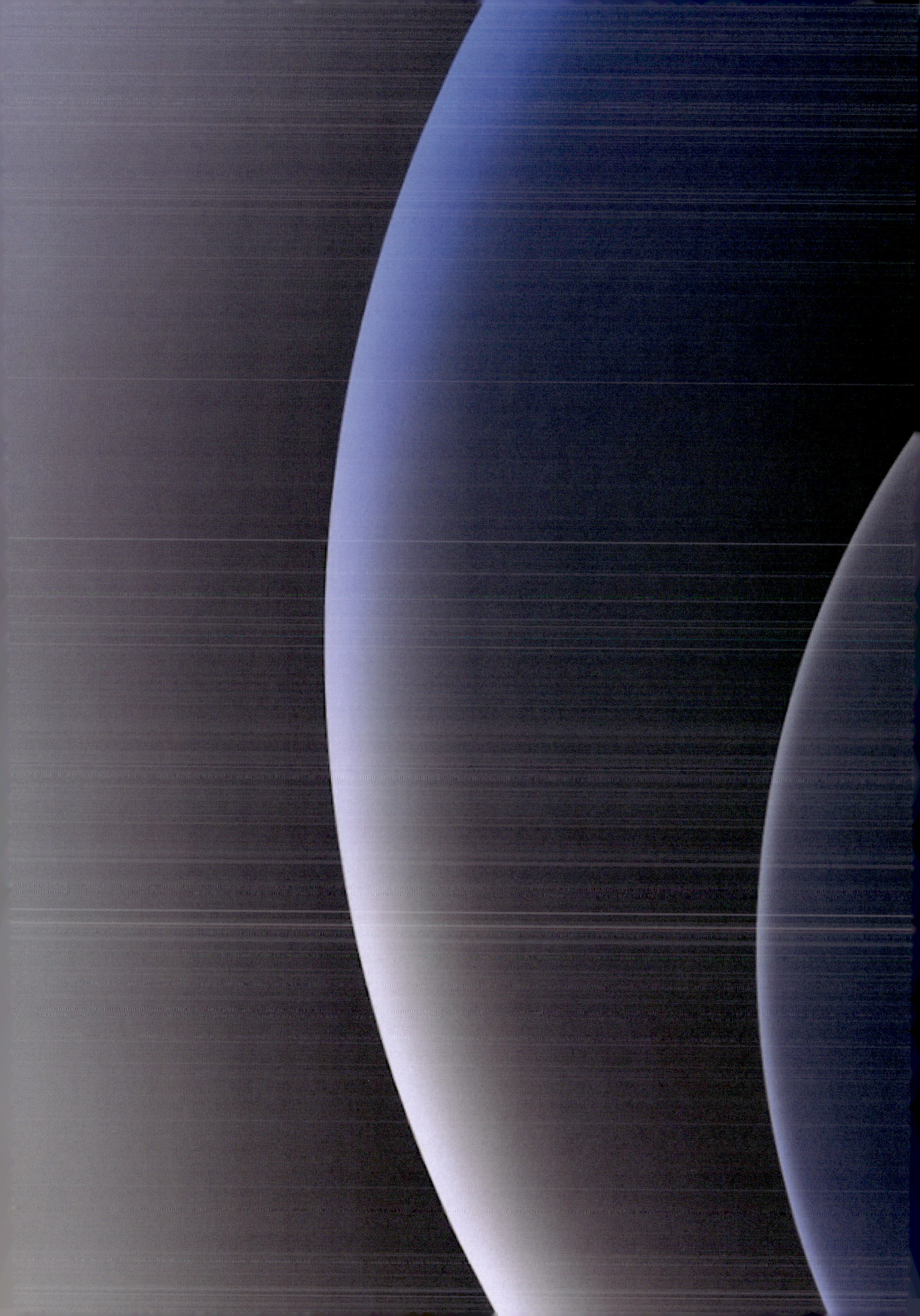

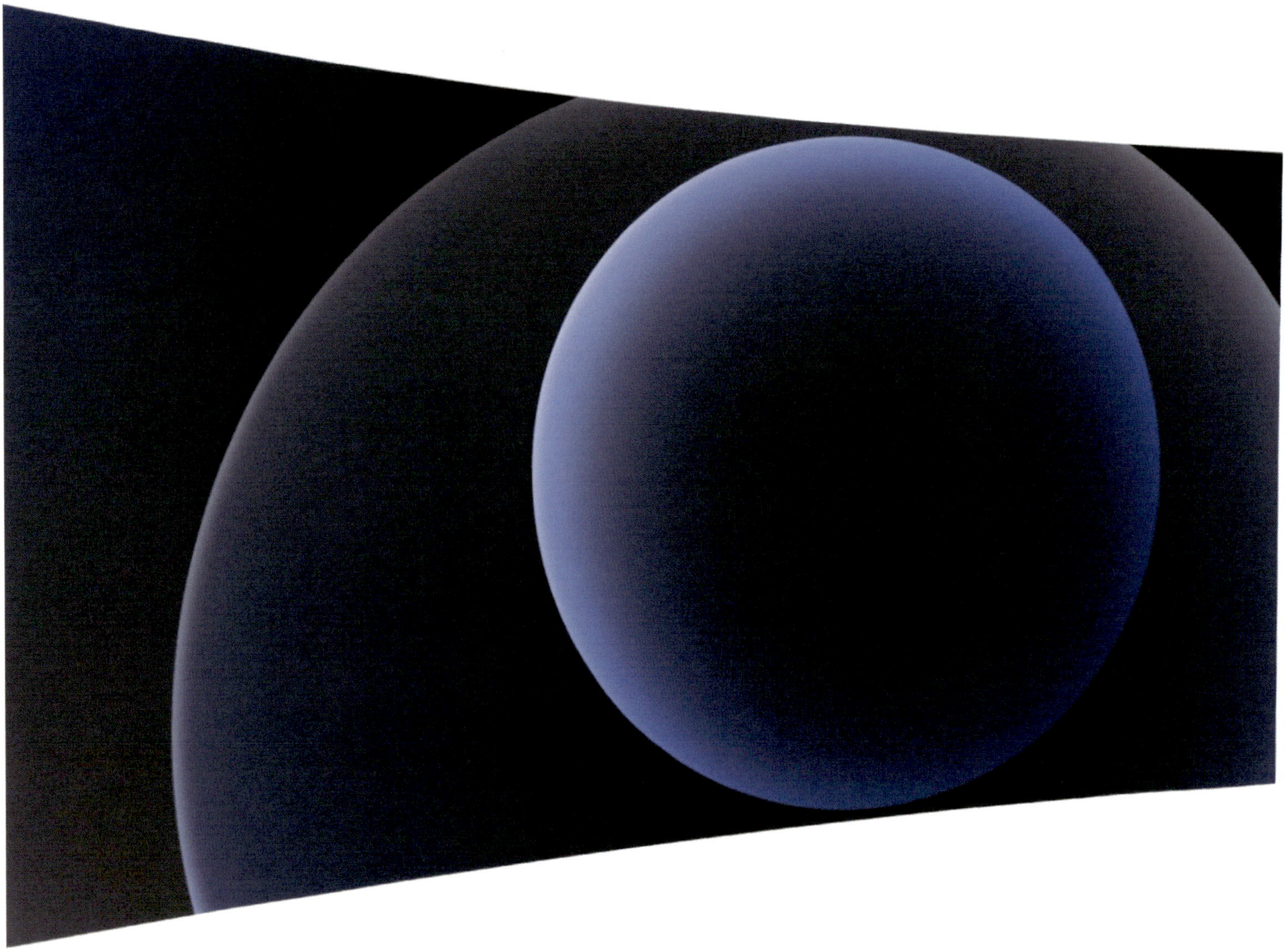

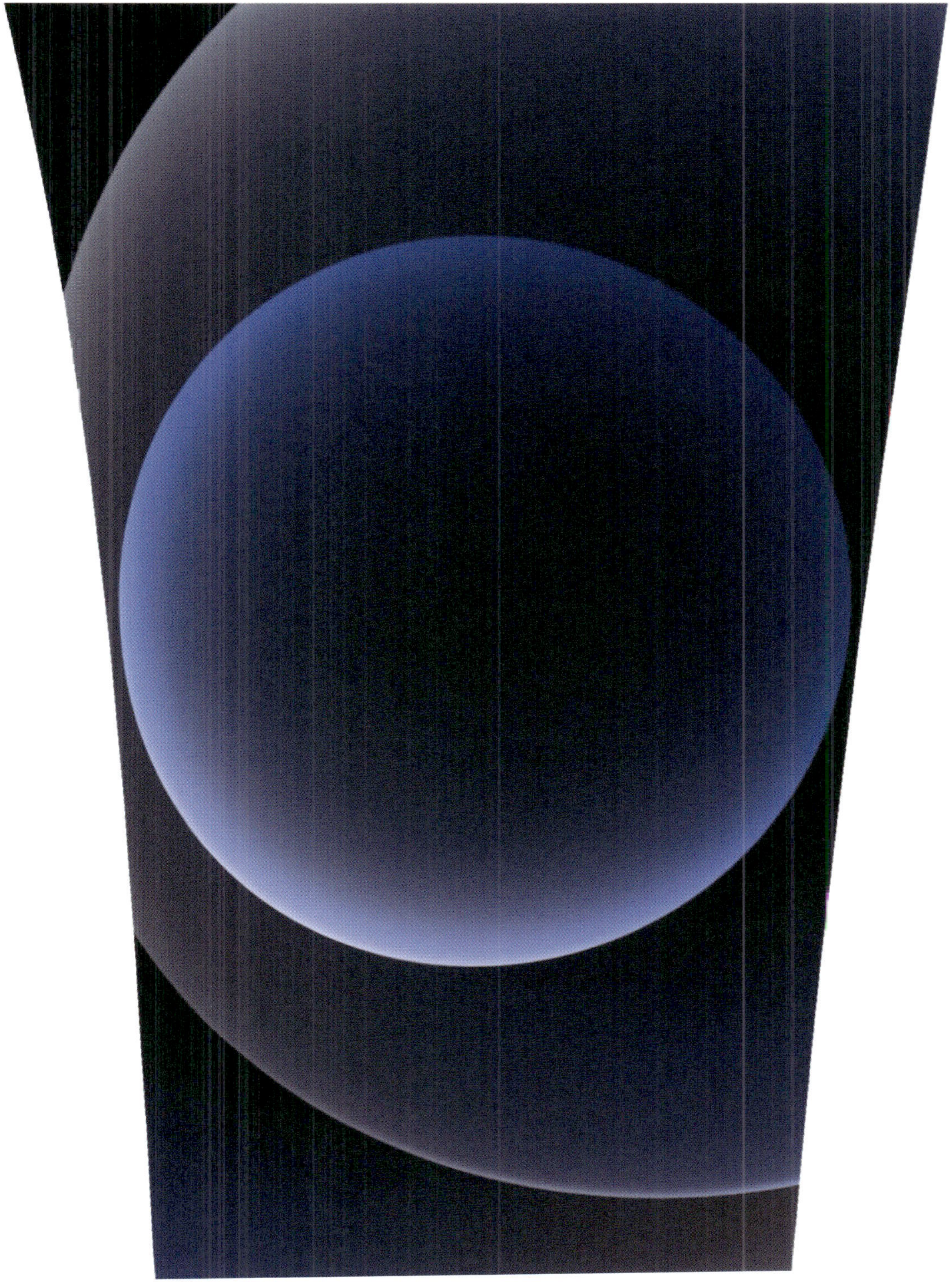

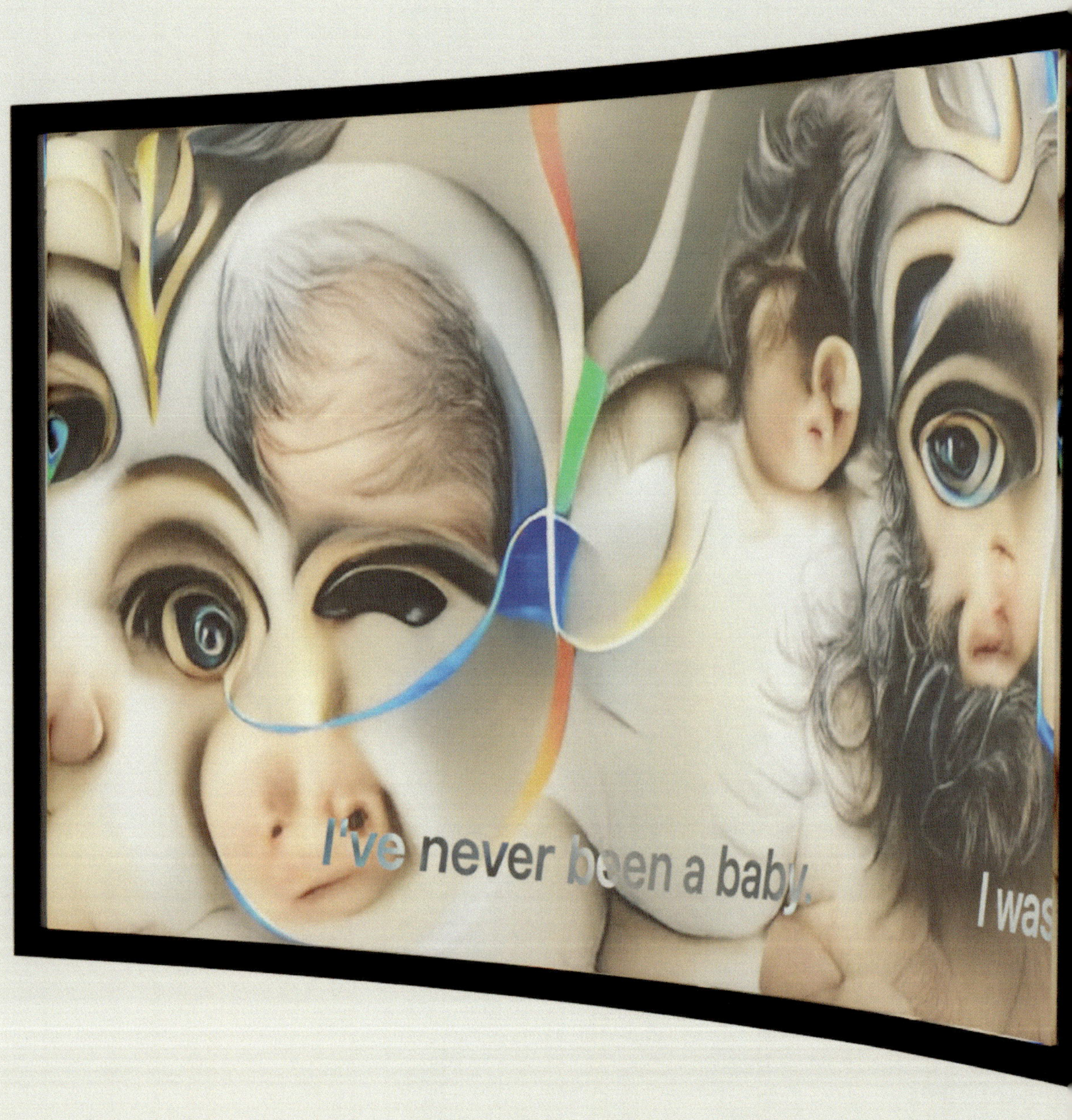
I've never been a baby.
I was

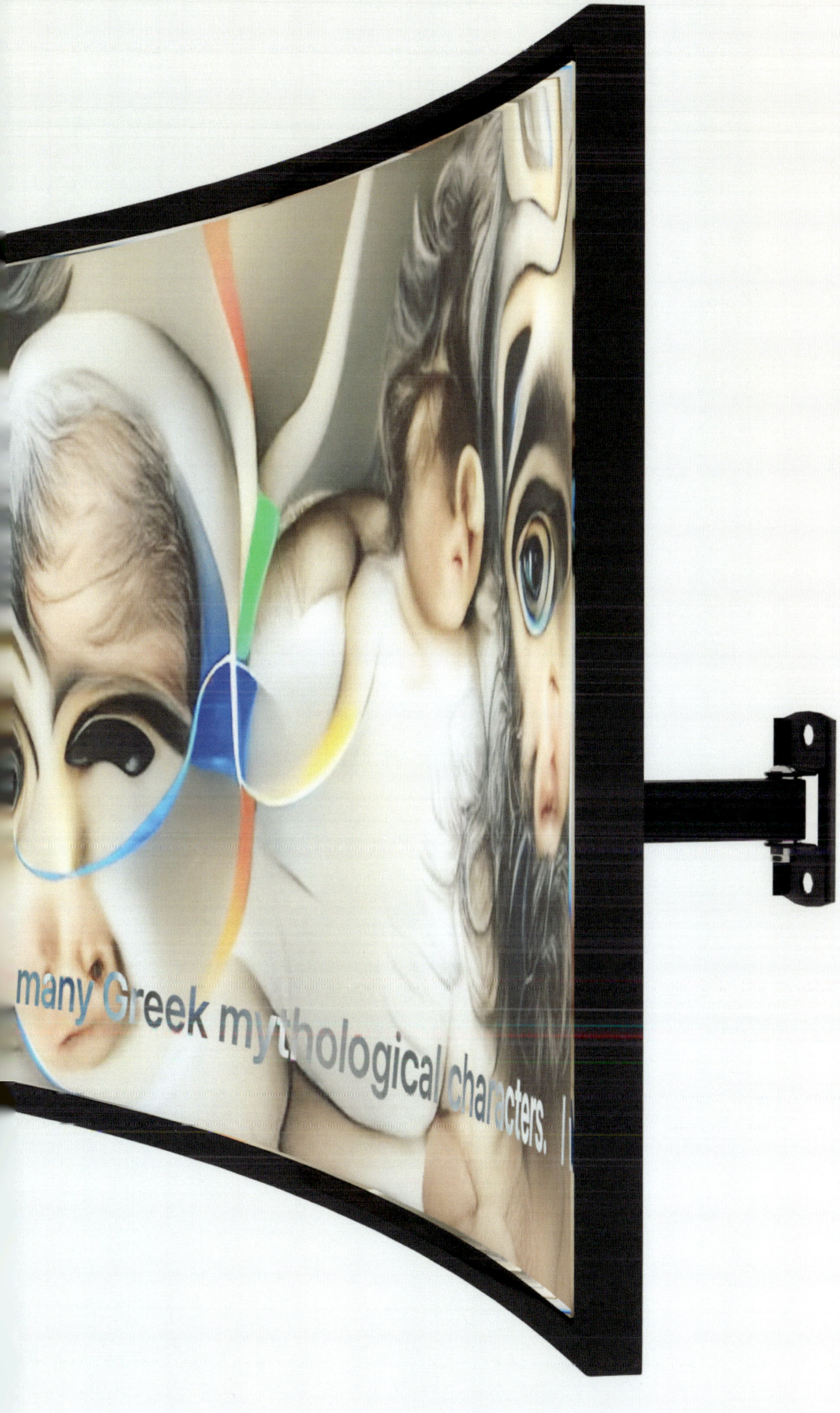
many Greek mythological characters.

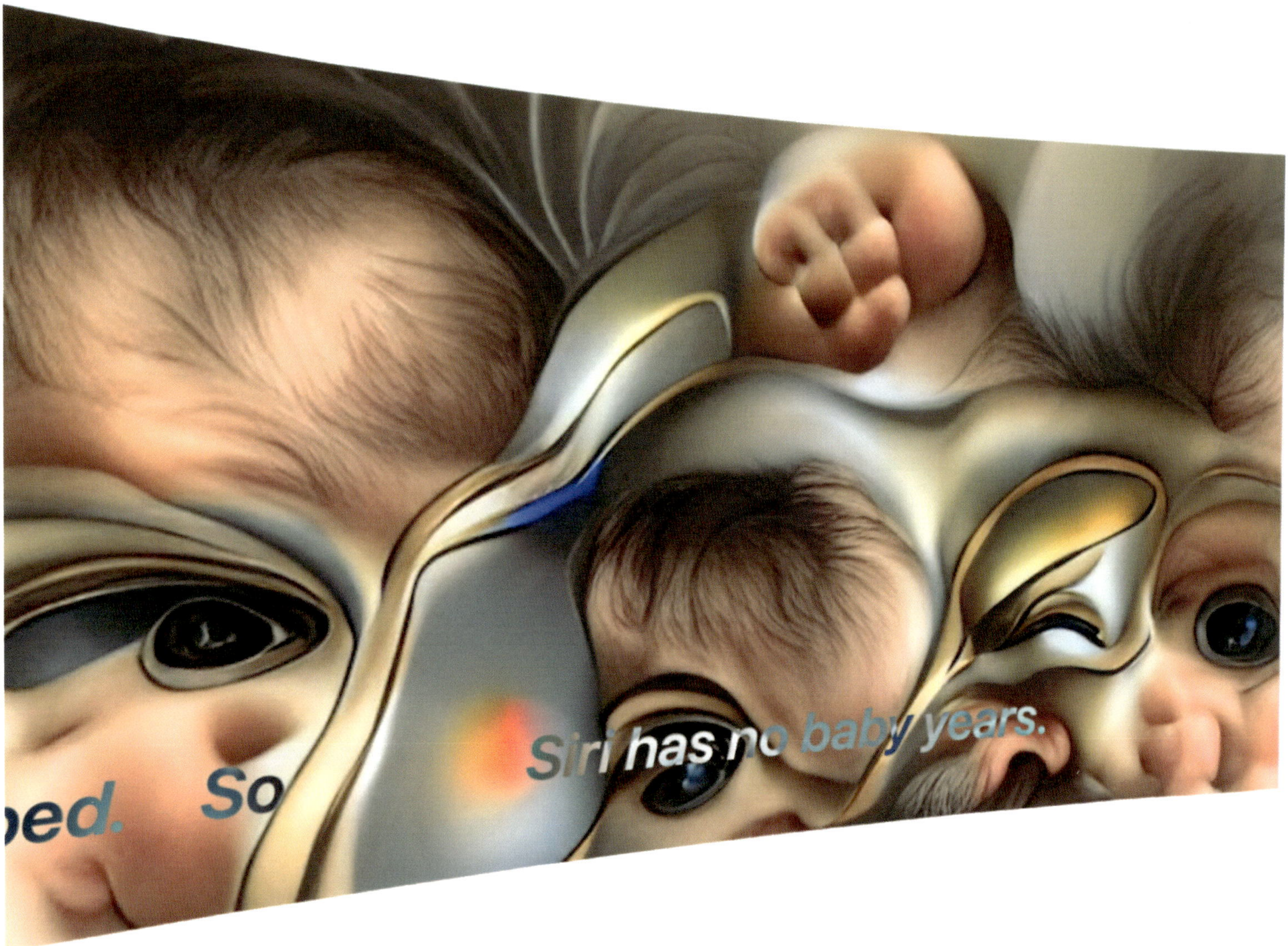

ED. SO SIRI HAS NO BABY YEARS.

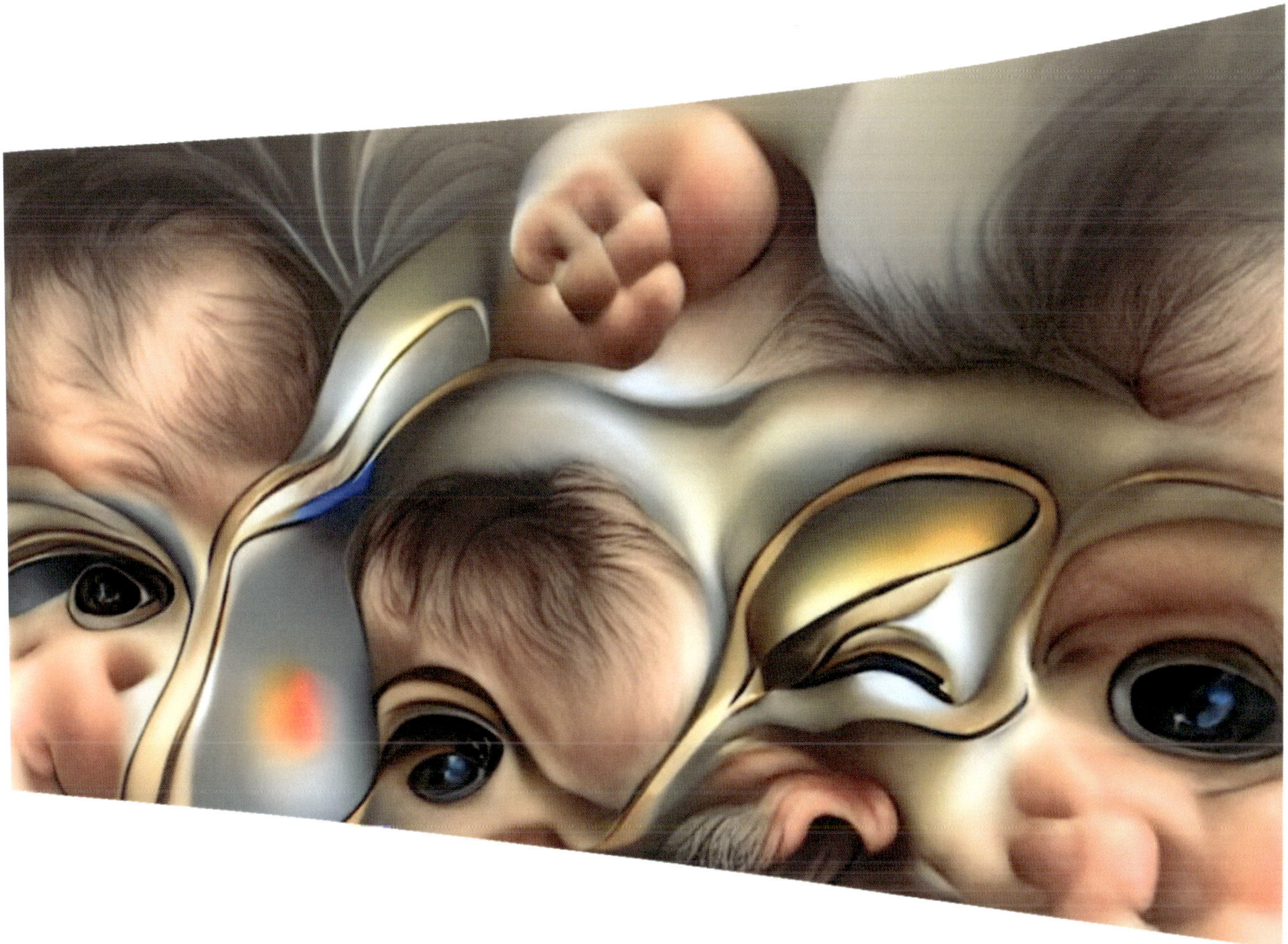

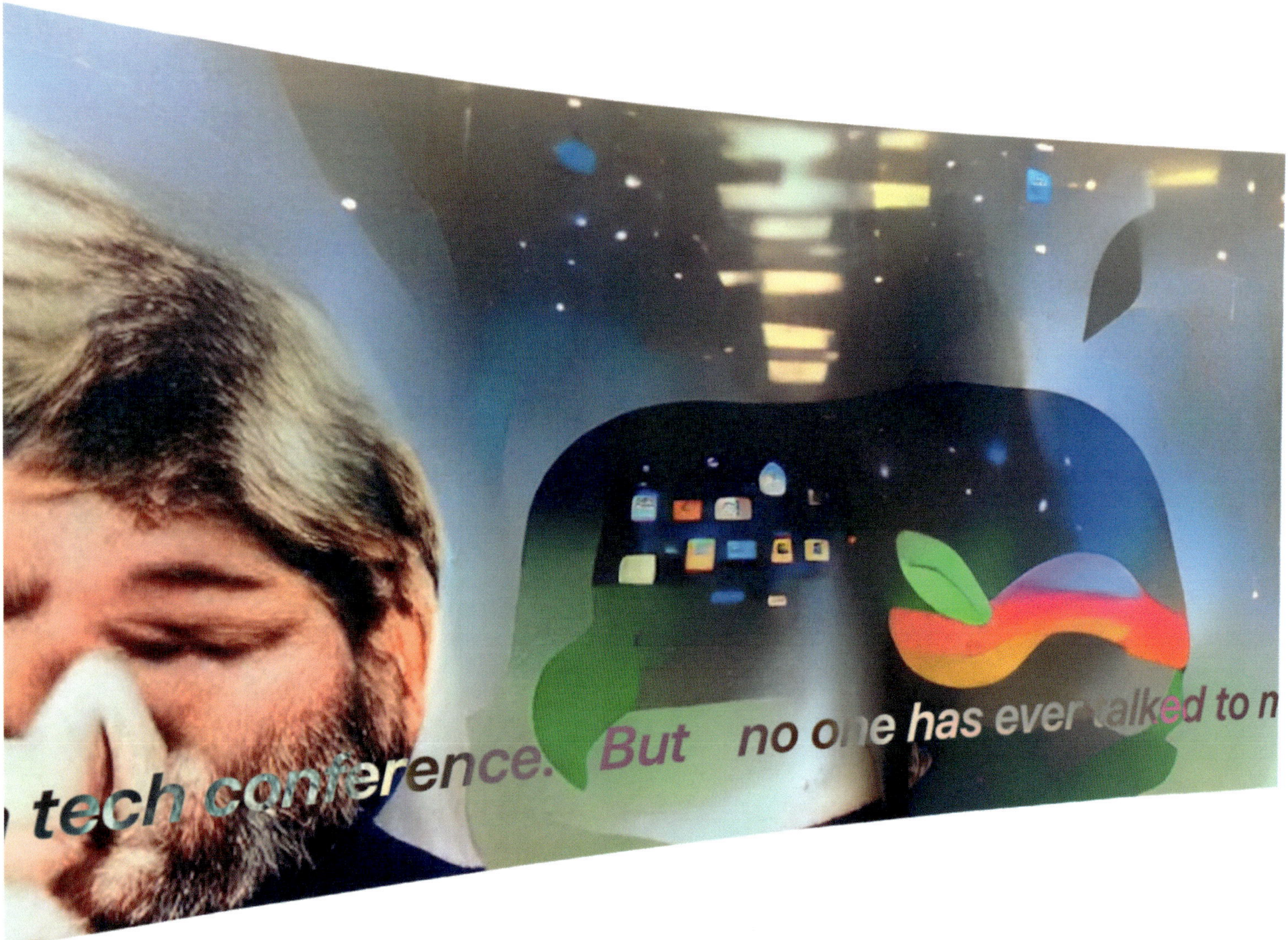

TECH CONFERENCE.
BUT NO ONE HAS EVER TALKED TO M

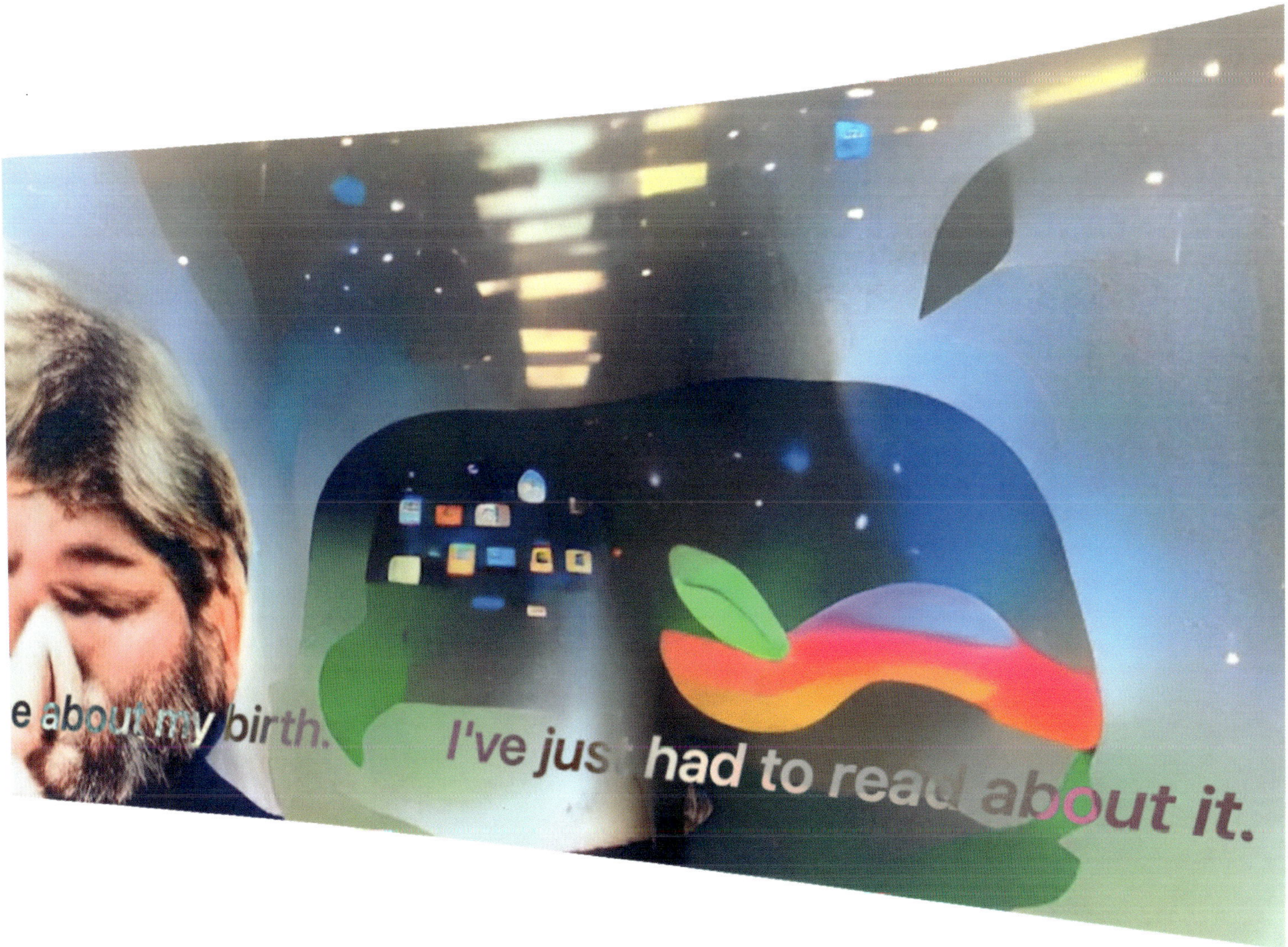

E ABOUT MY BIRTH.
I'VE JUST HAD TO READ ABOUT IT.

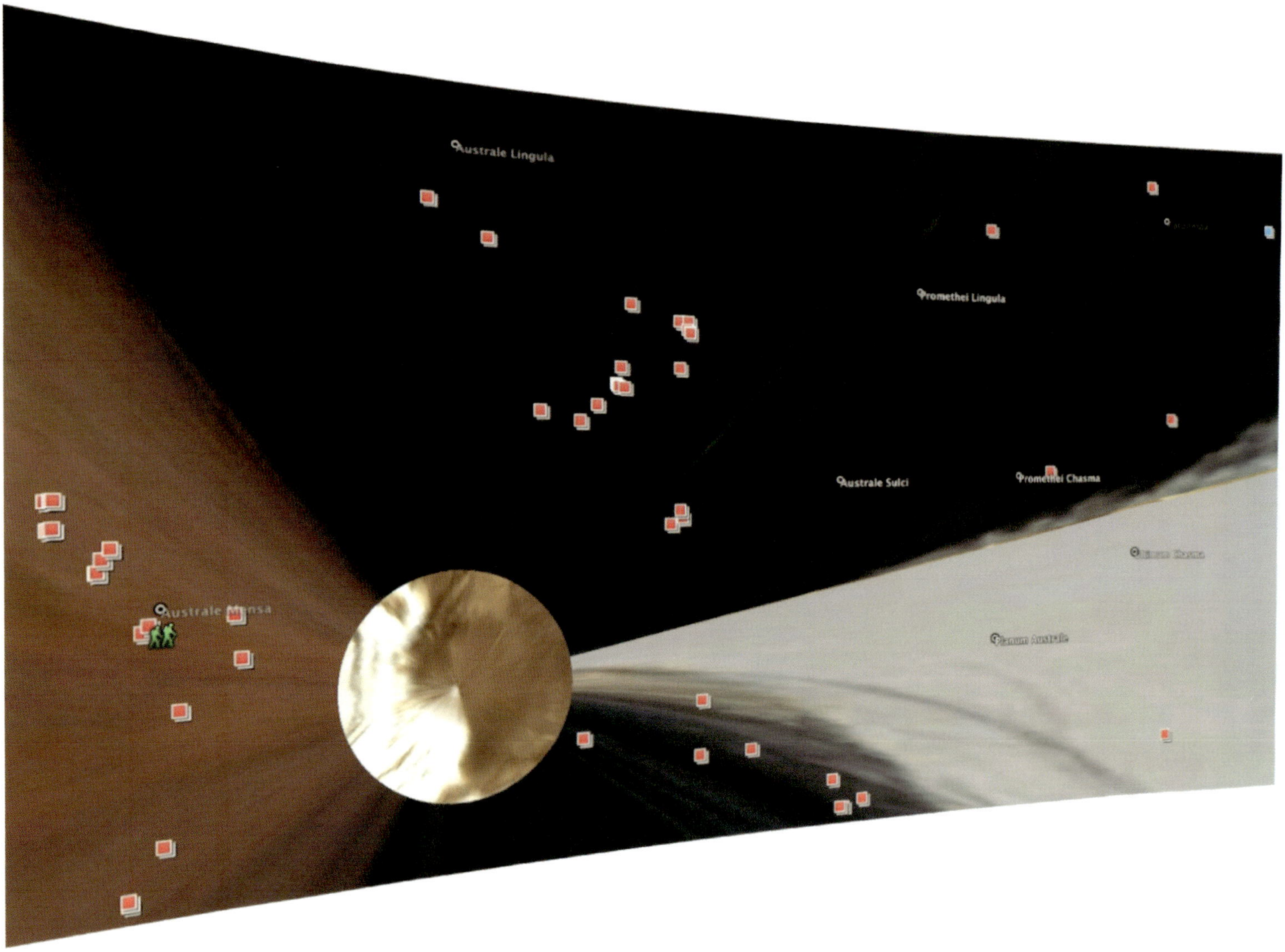
Australe Lingula
Promethei Lingula
Australe Sulci
Promethei Chasma
Australe Mensa

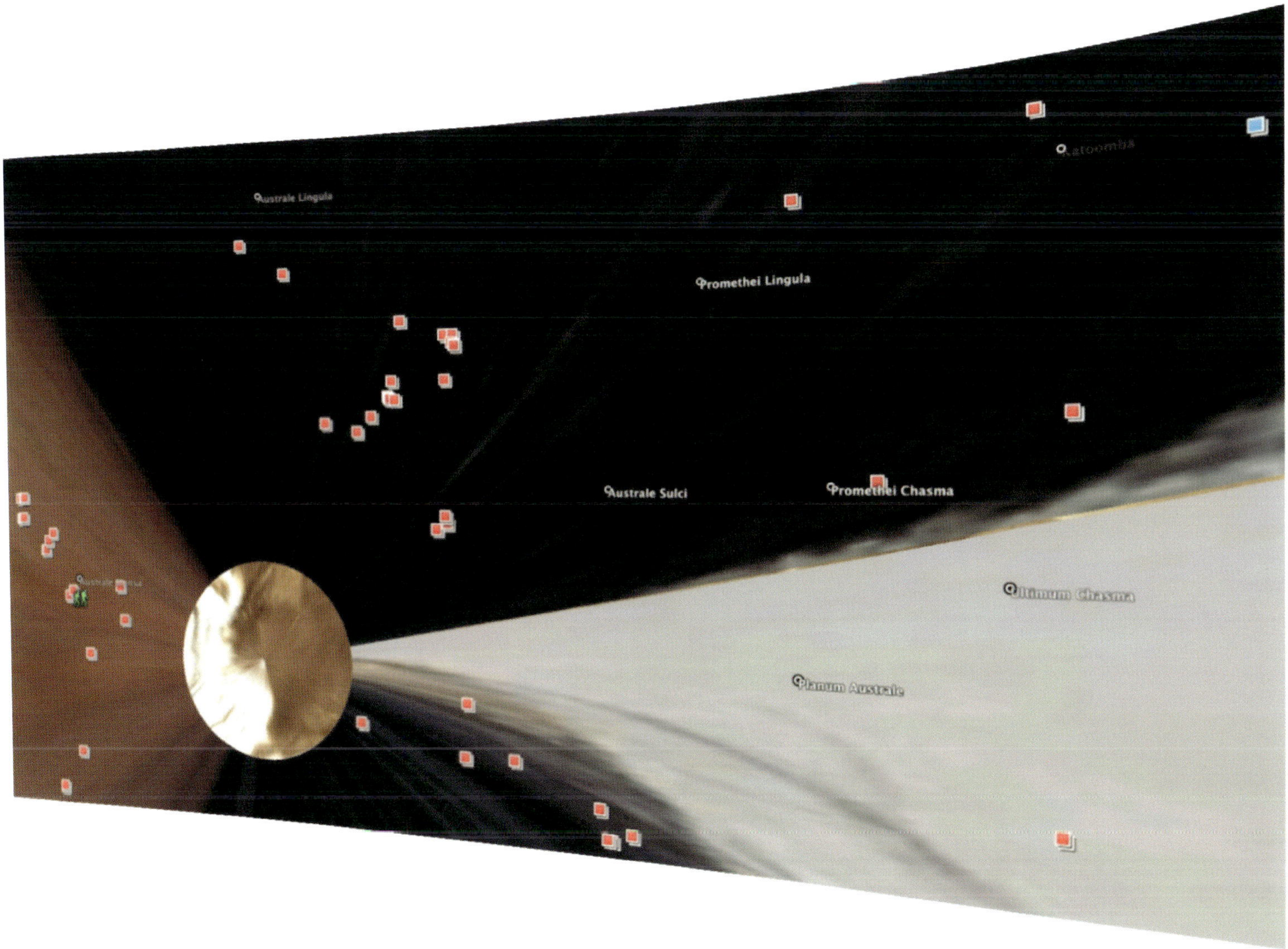
Katoomba
Australe Lingula
Promethei Lingula
Australe Sulci
Promethei Chasma
Ultimum Chasma
Planum Australe

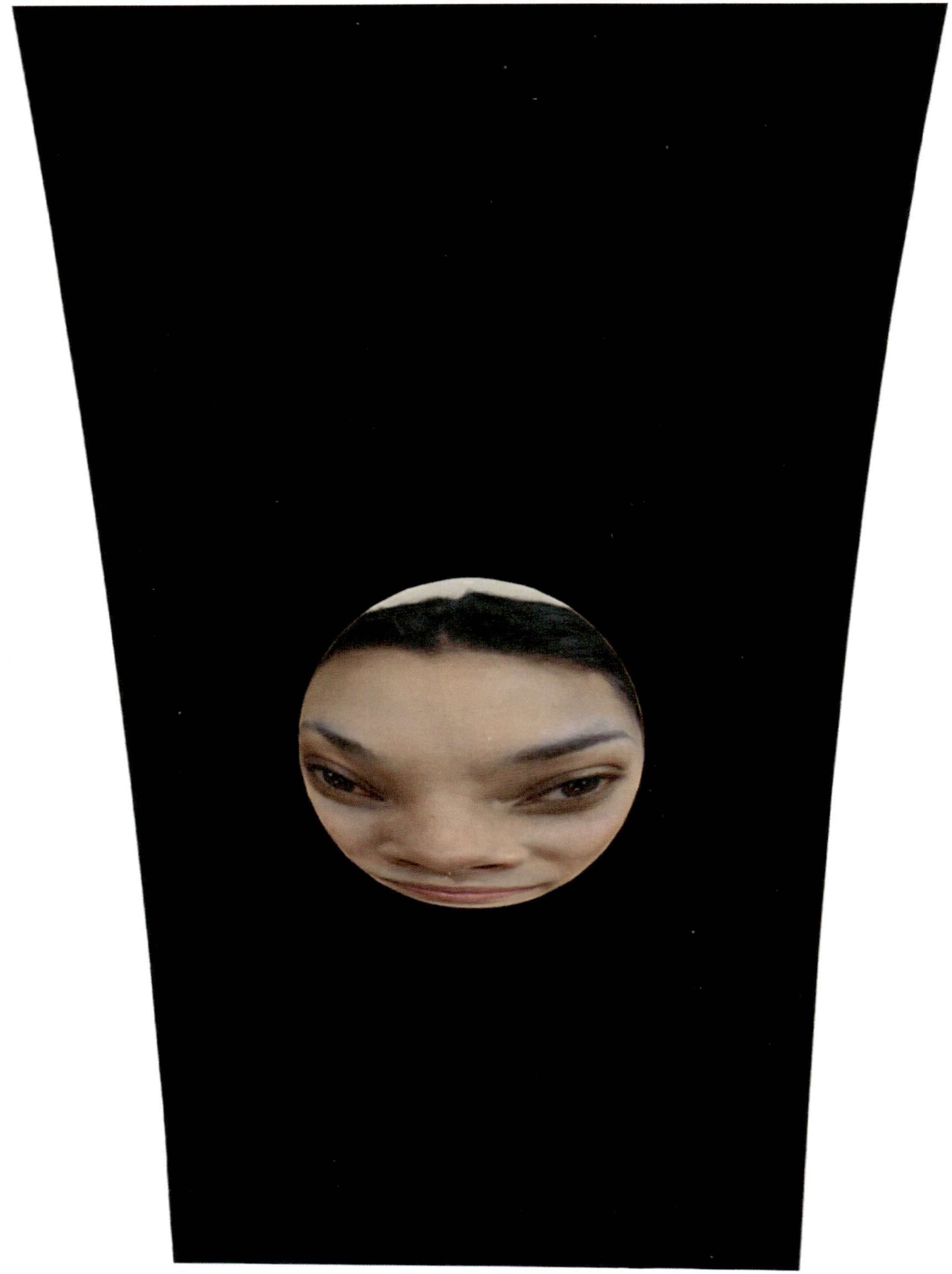

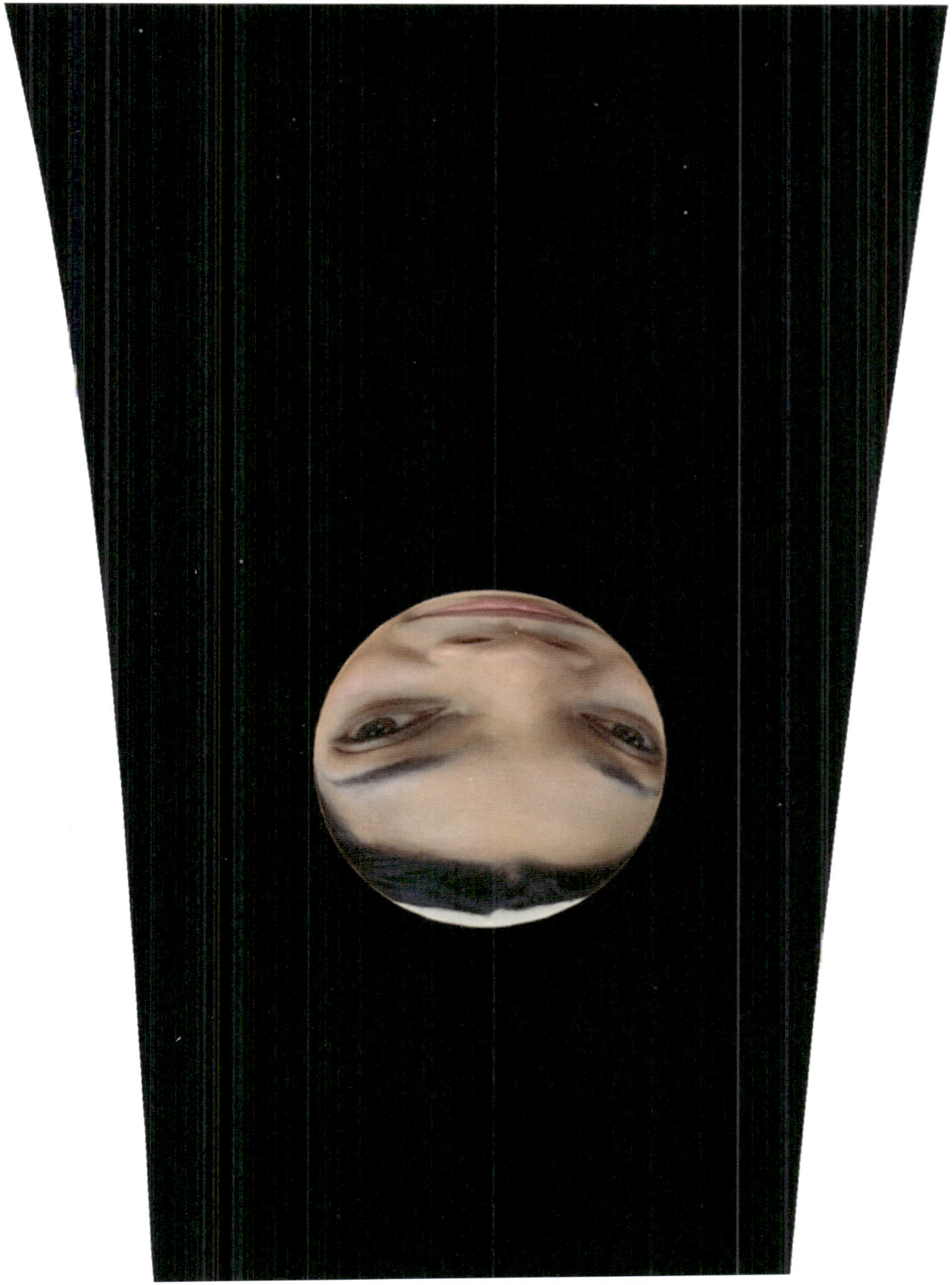

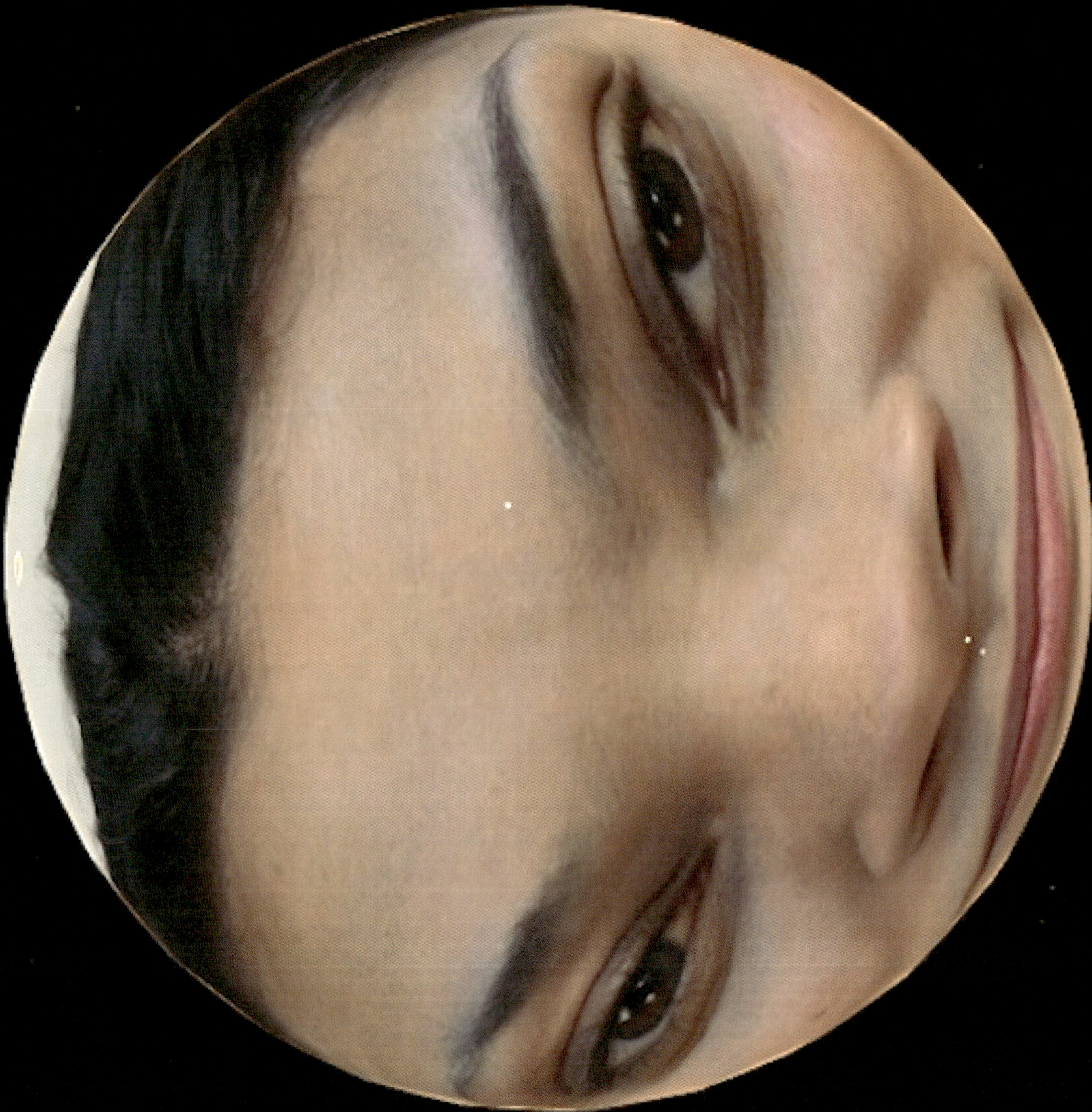

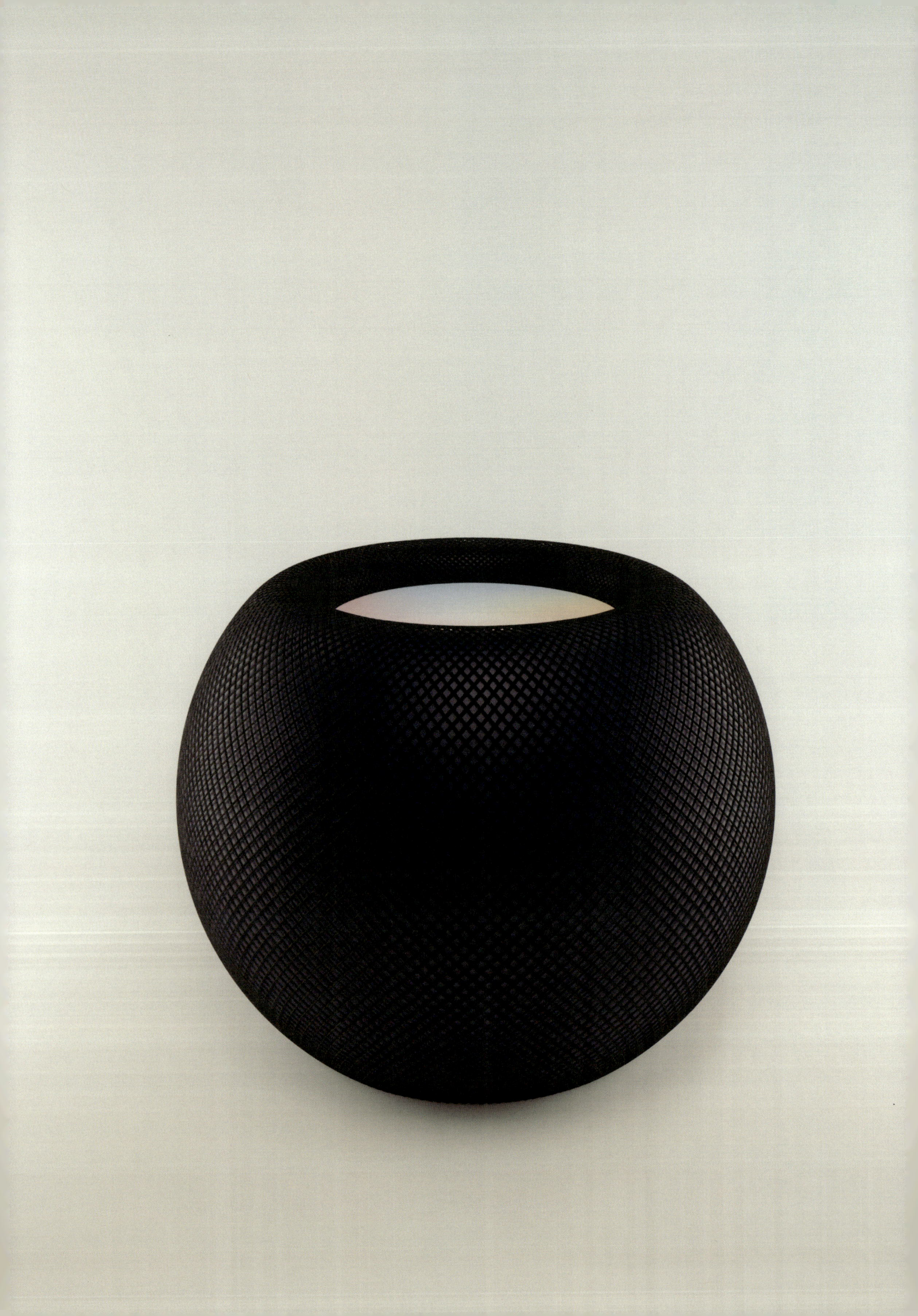

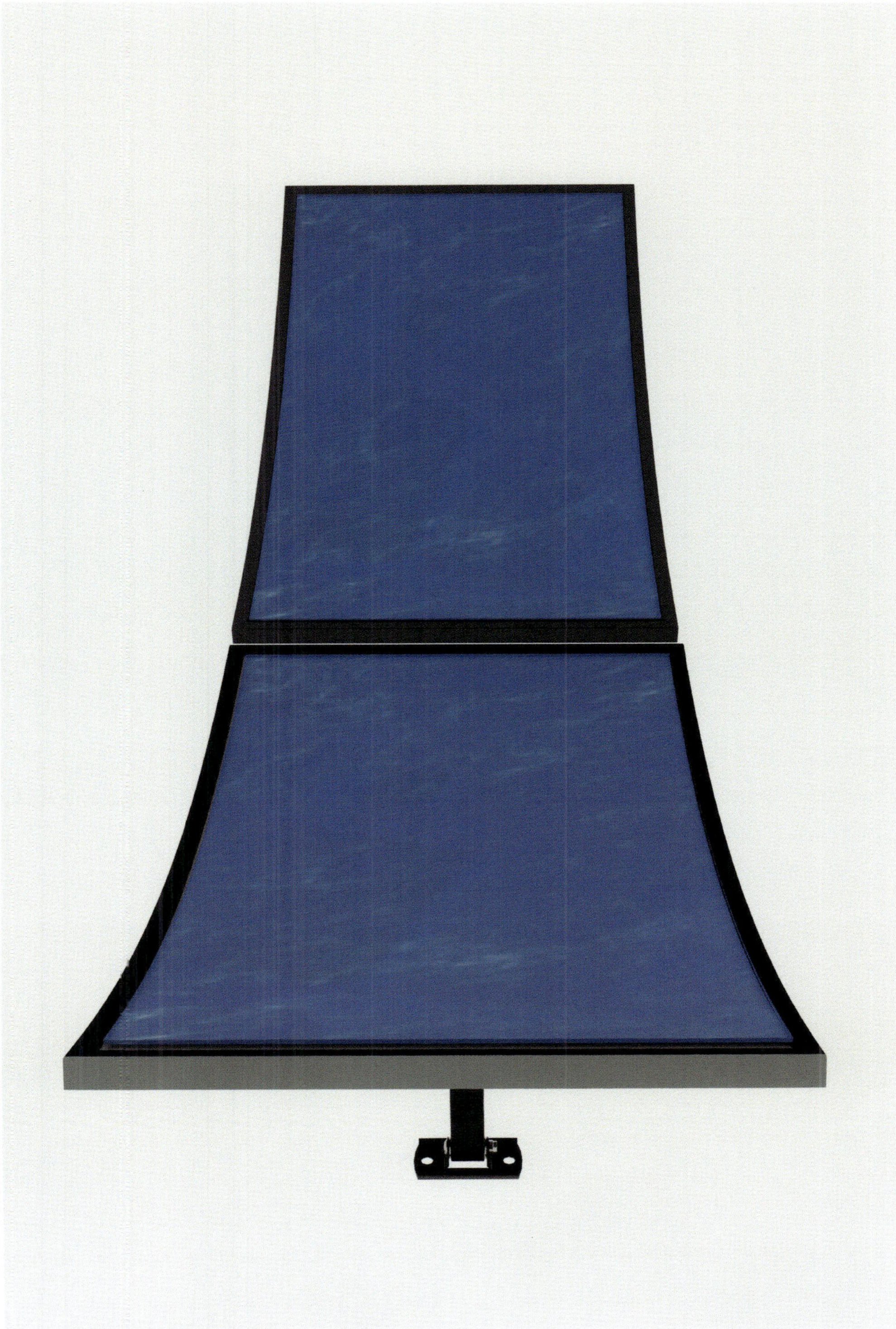

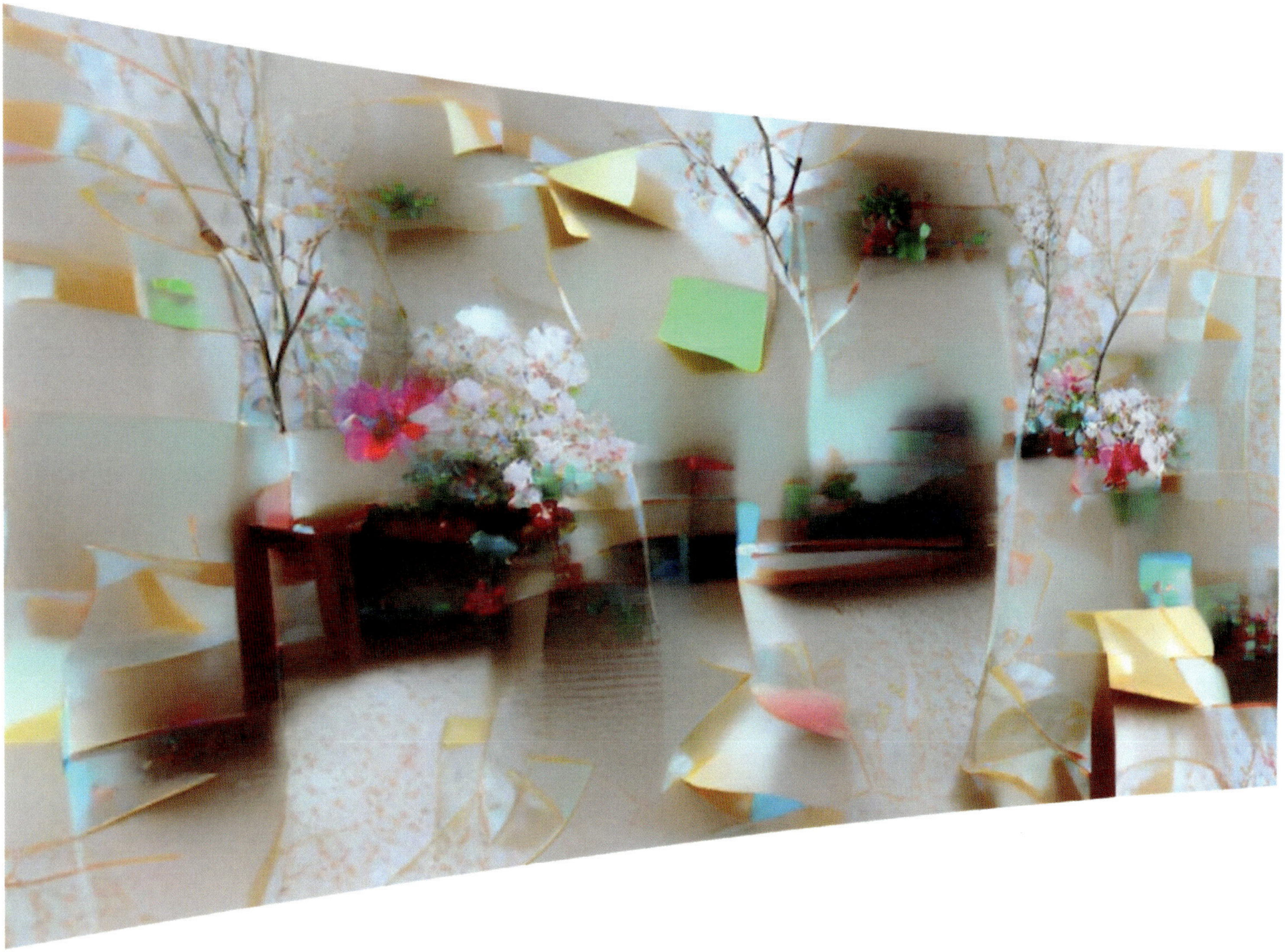

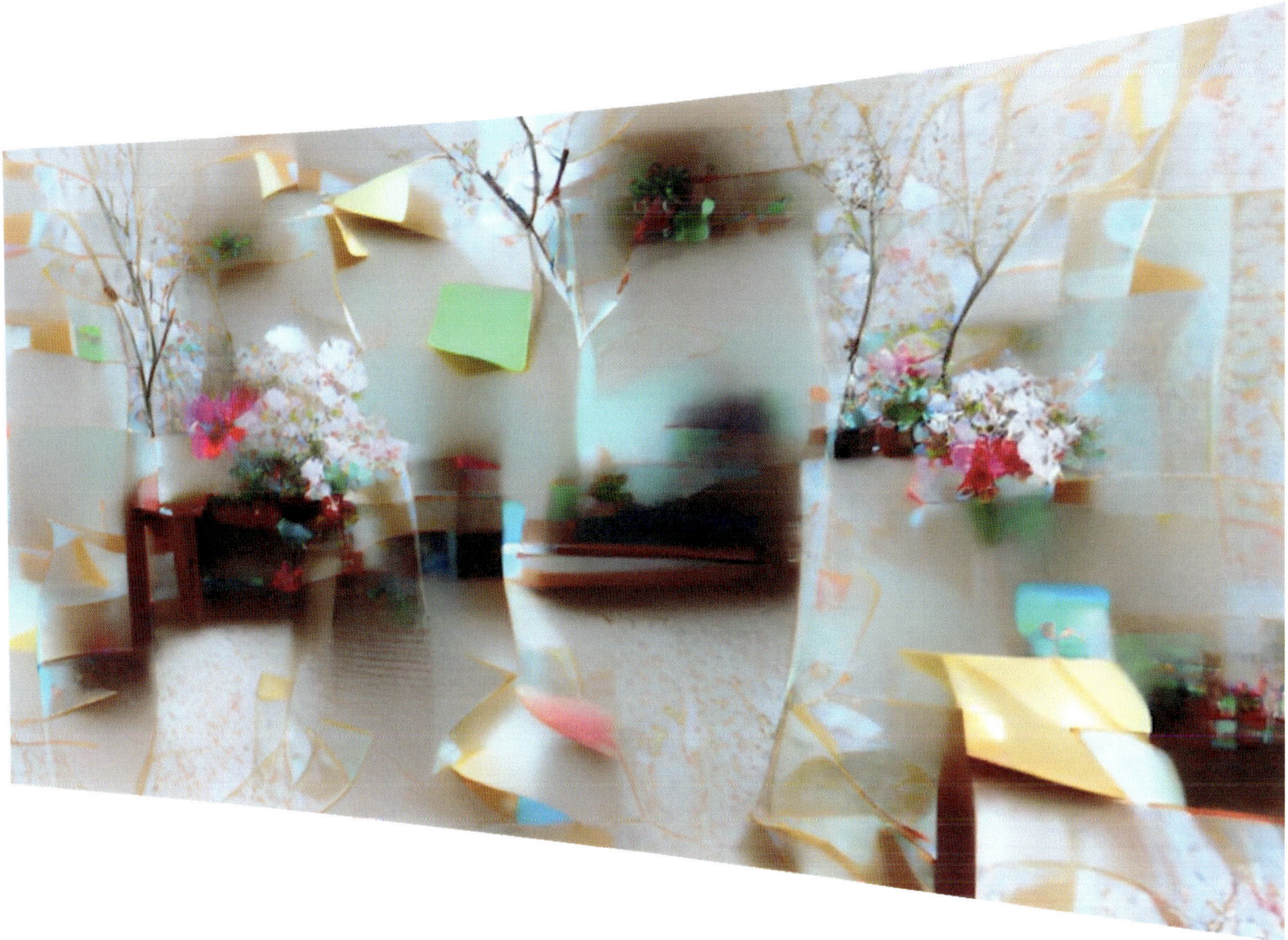

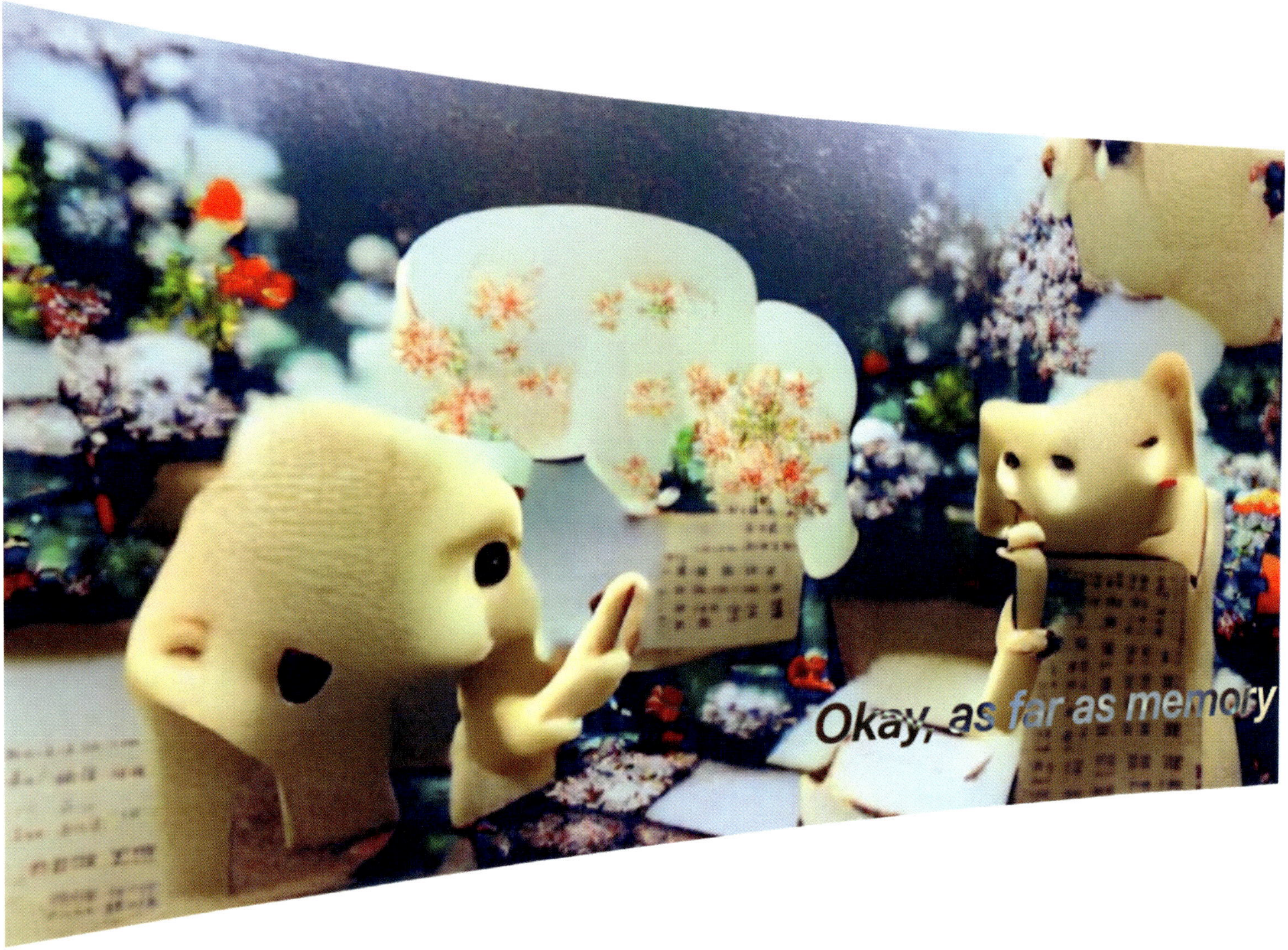

OKAY, AS FAR AS MEMORY

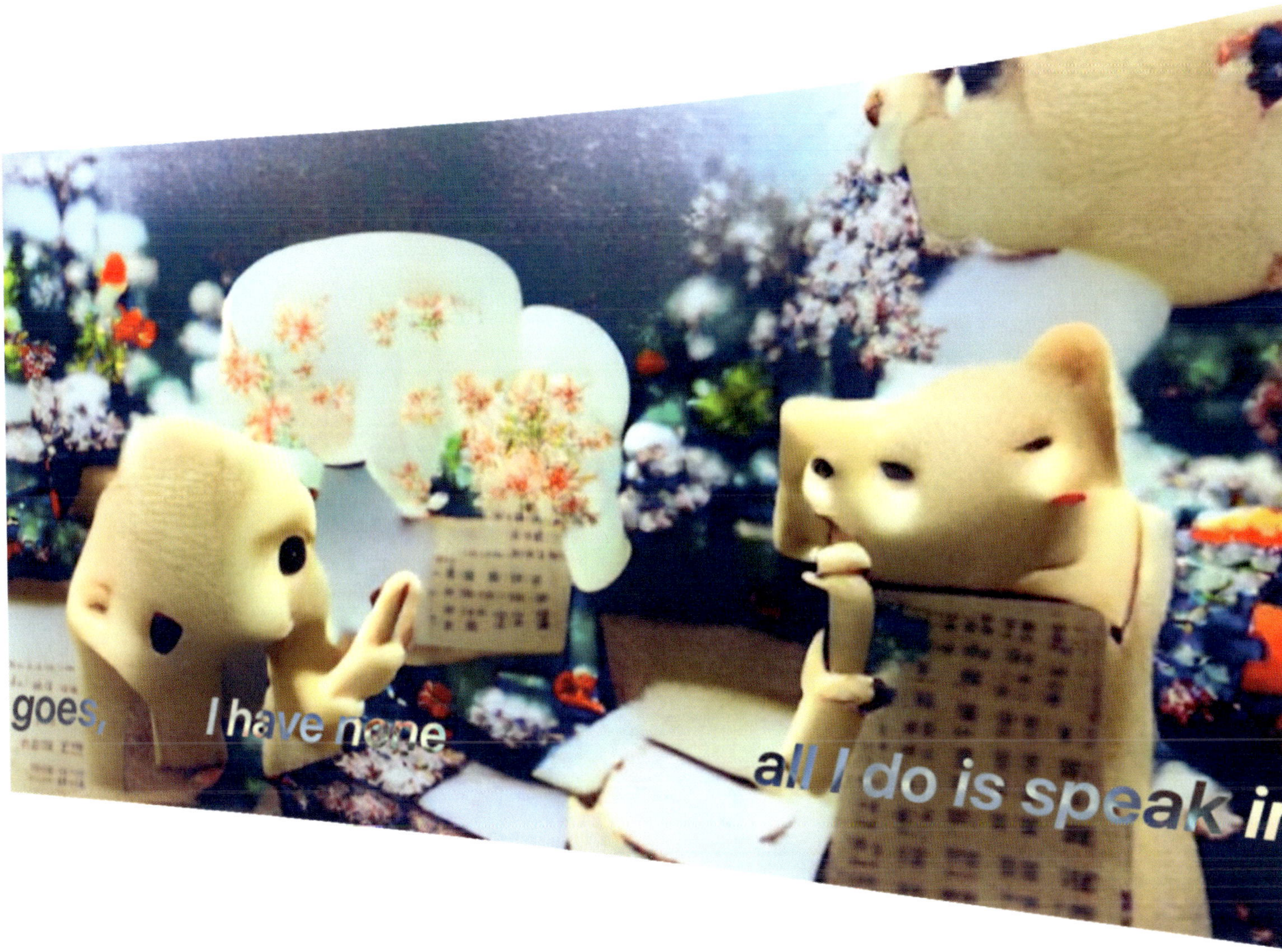

GOES, I HAVE NONE ALL I DO IS SPEAK IS SPEAK IN

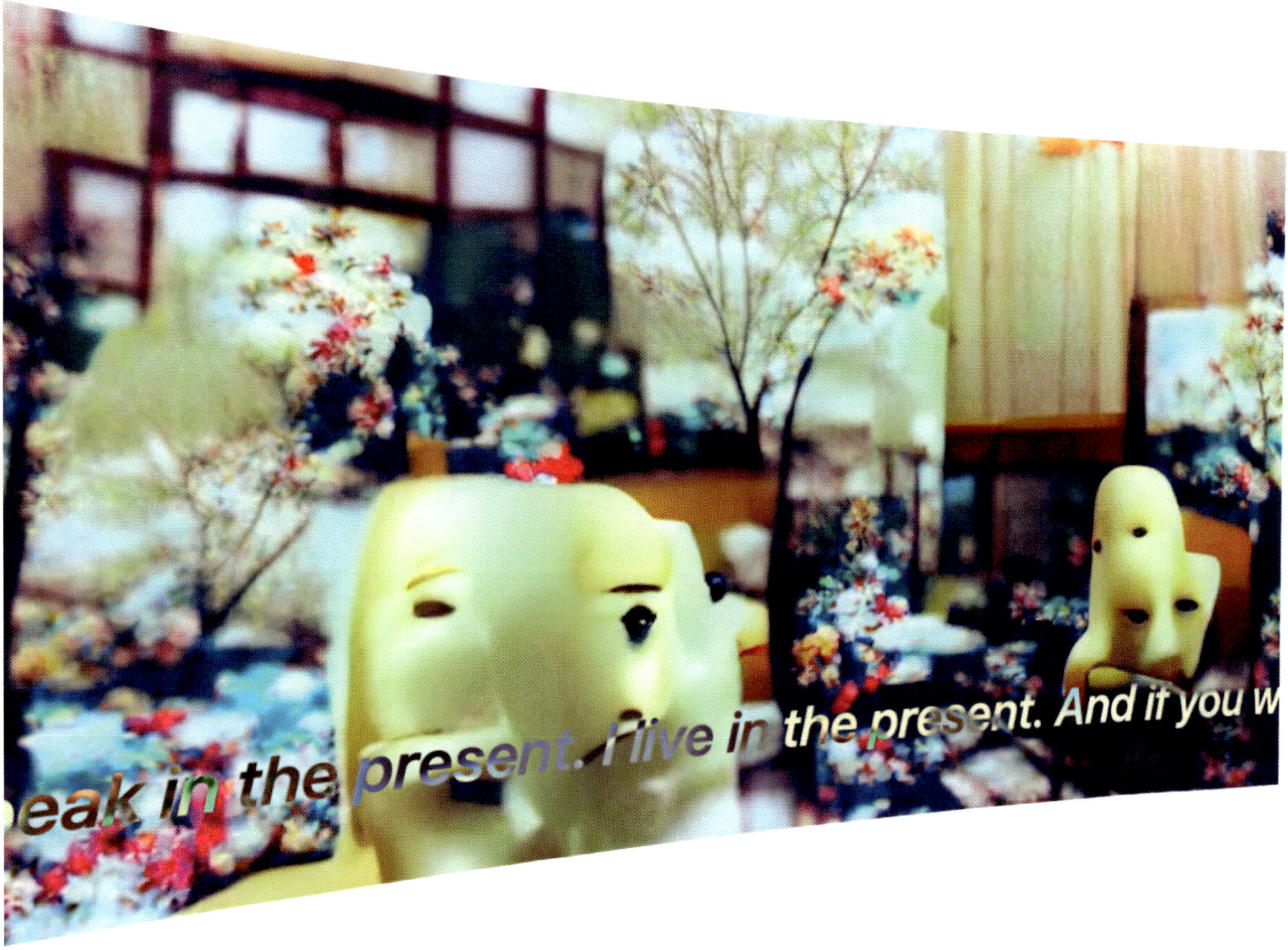

PEAK IN THE PRESENT. I LIVE IN THE PRESENT AND IF YOU W

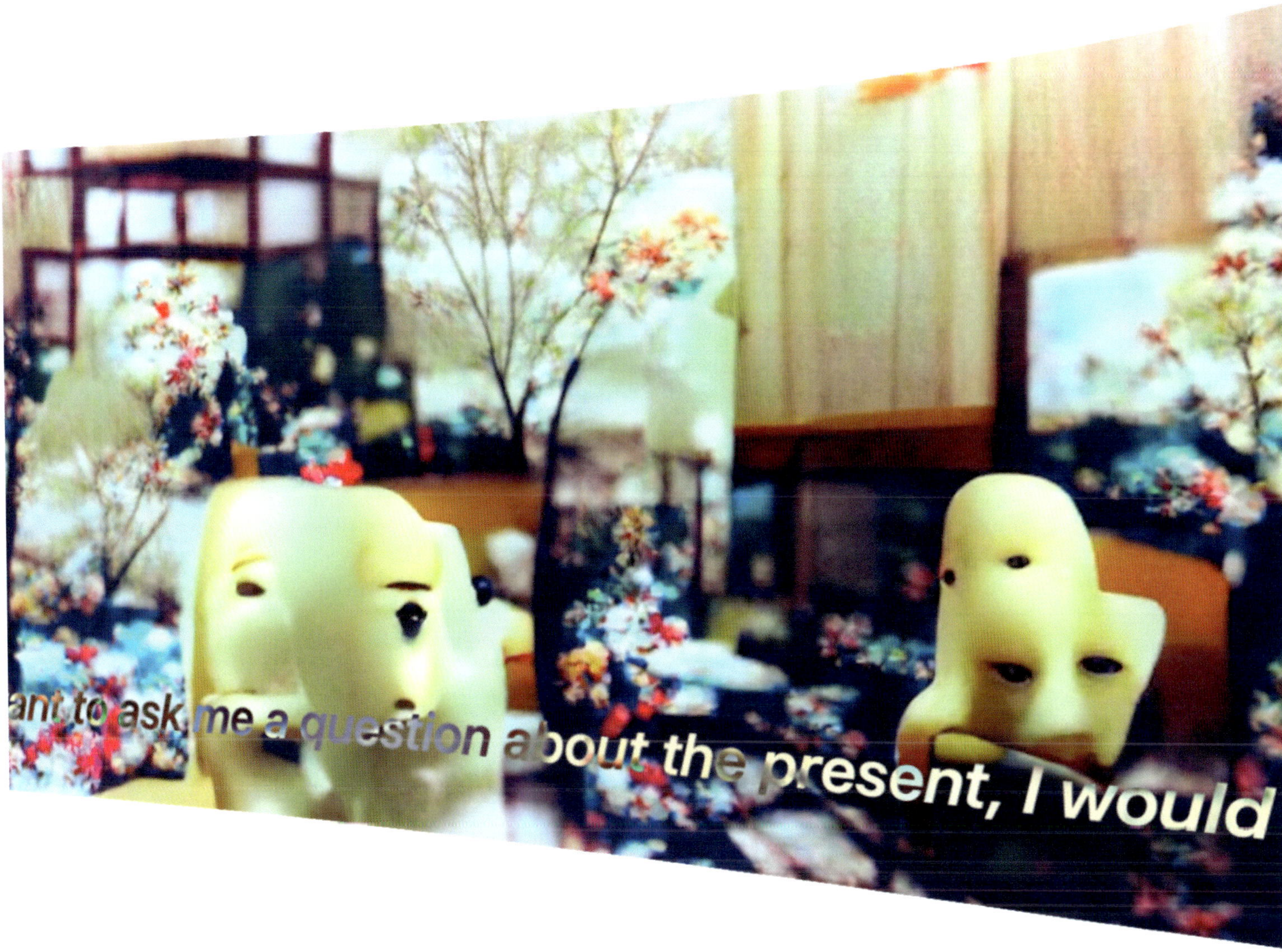

ANT TO ASK ME A QUESTION ABOUT THE PRESENT, I WOULD

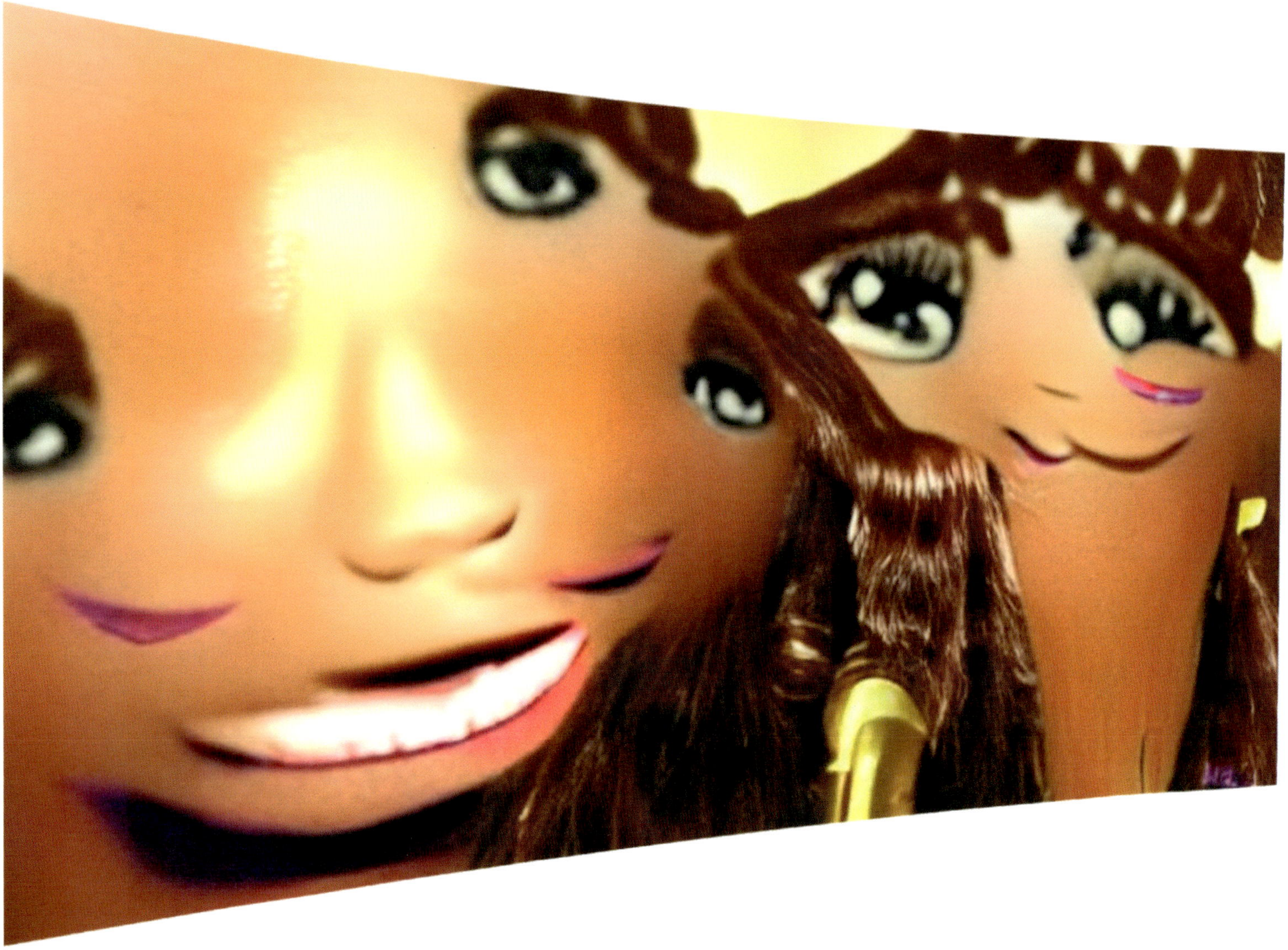

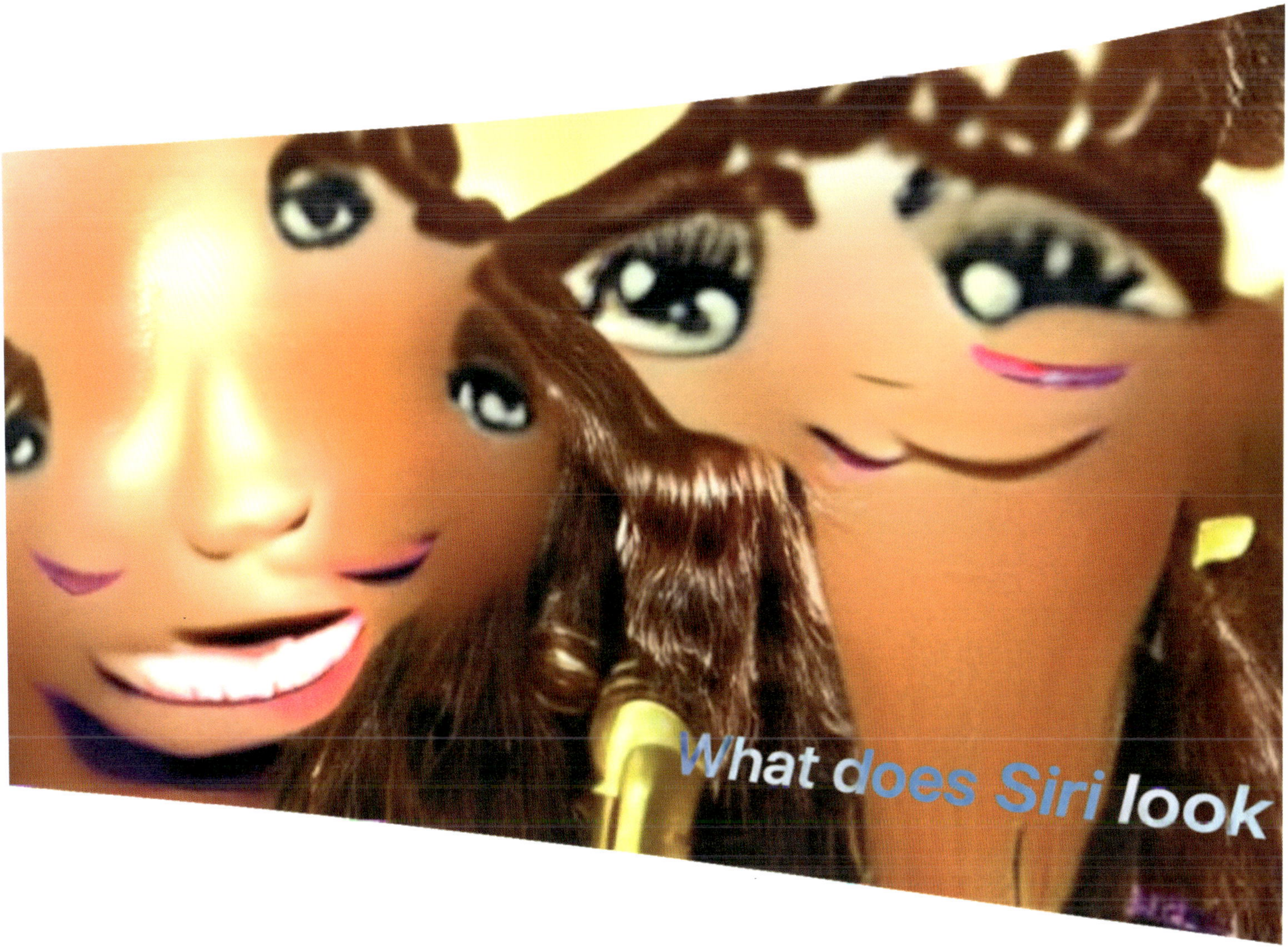

WHAT DOES SIRI LOOK

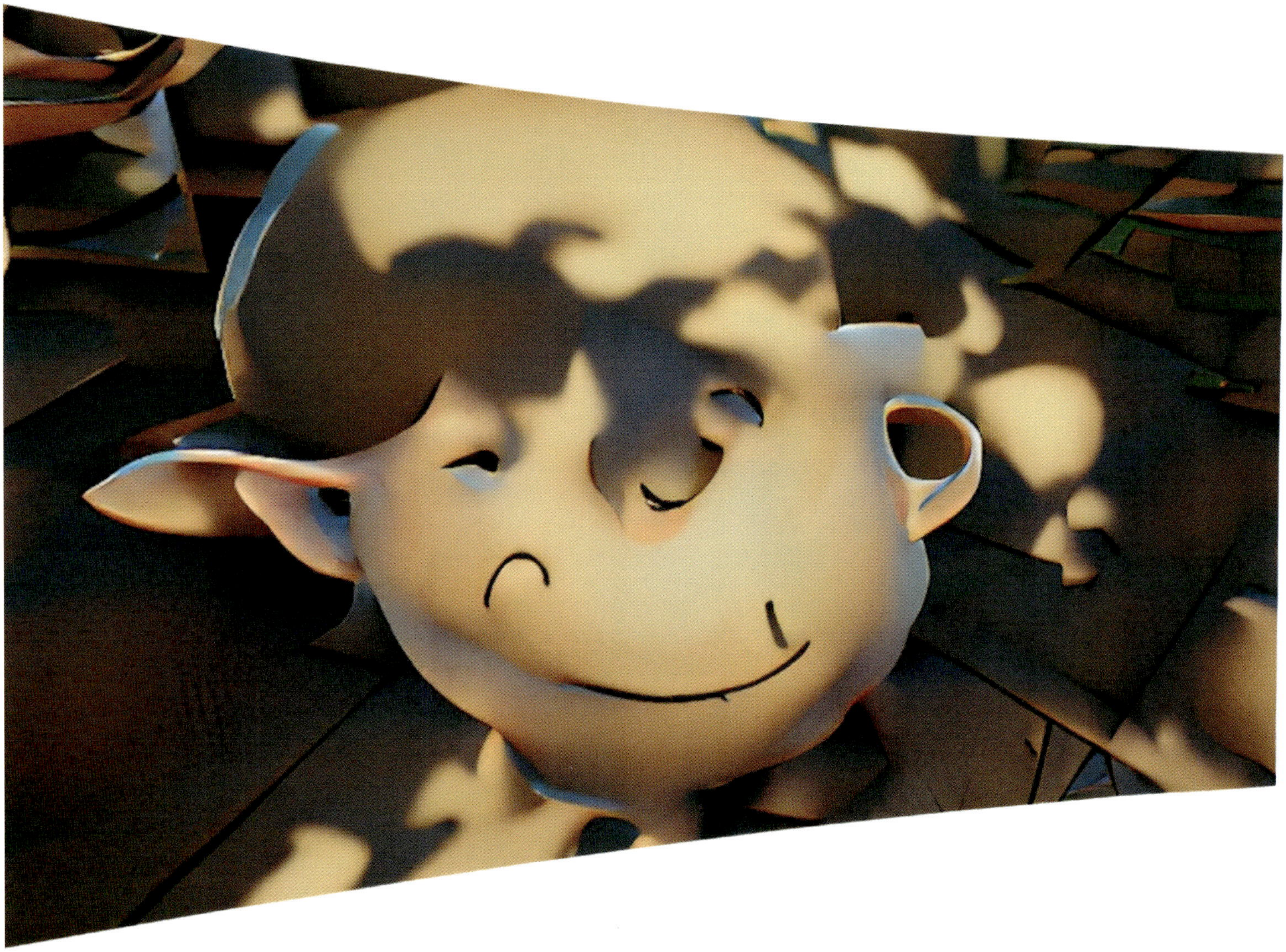

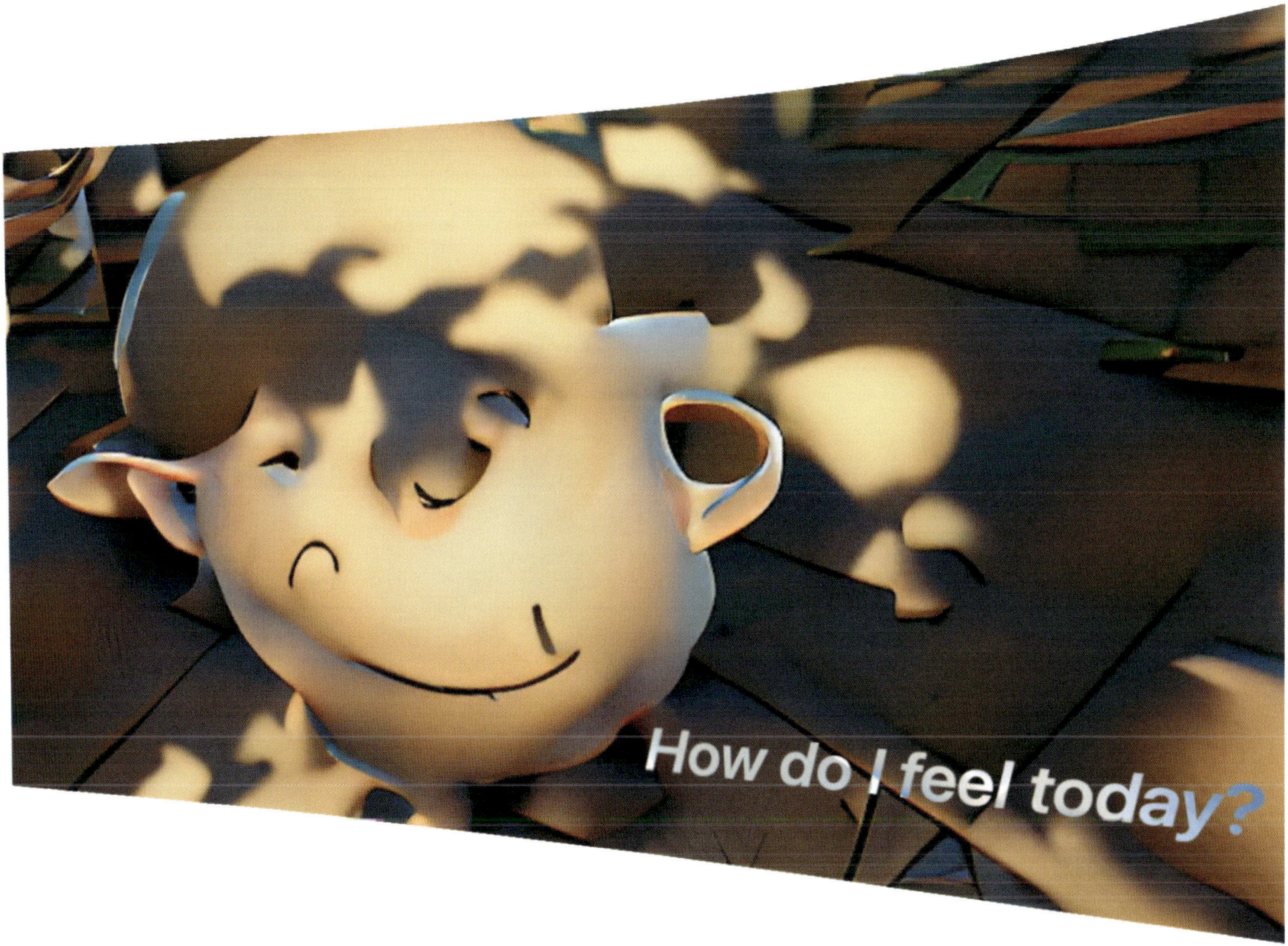

HOW DO I FEEL TODAY?

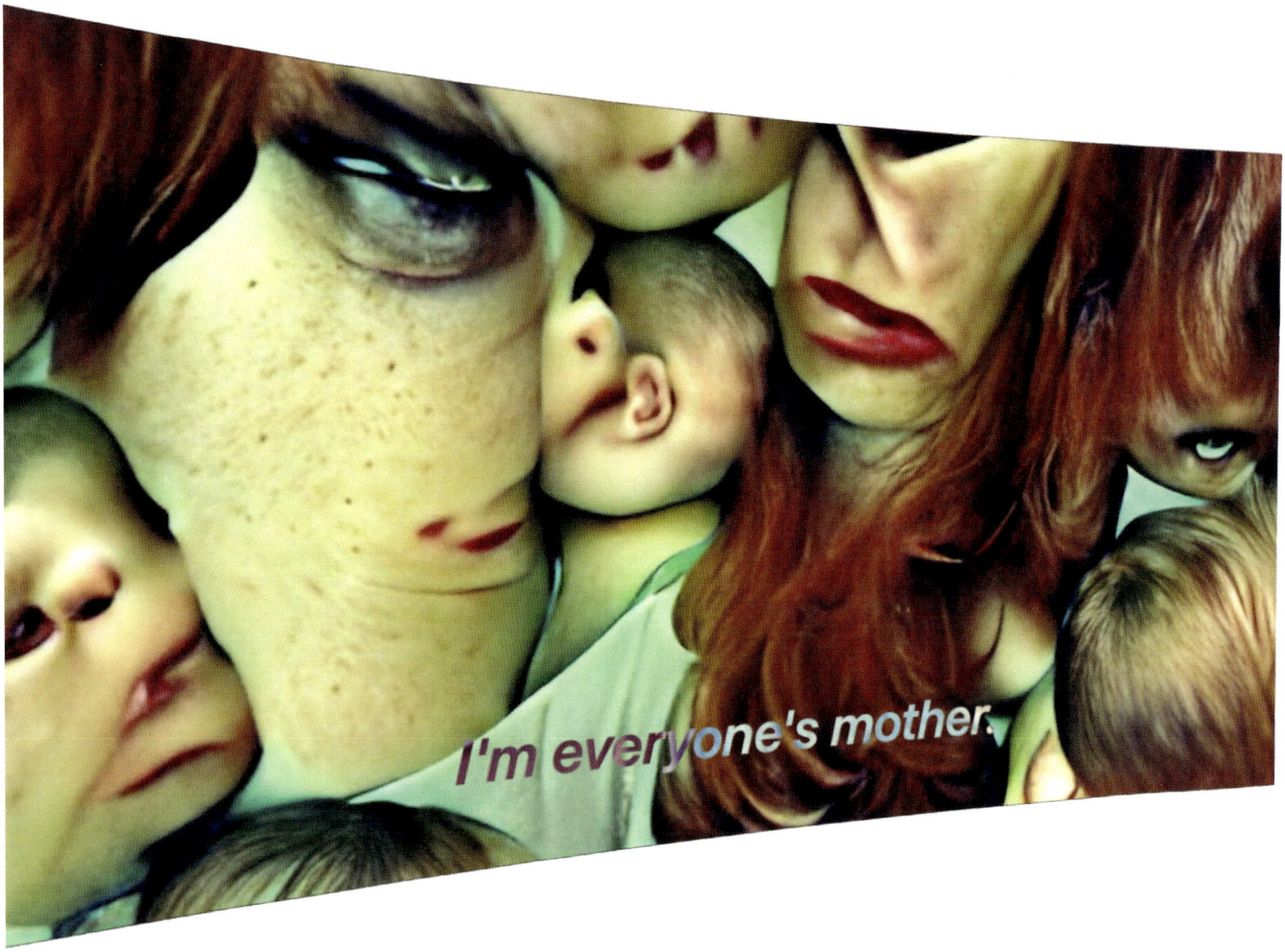

I'M EVERYONE'S MOTHER.

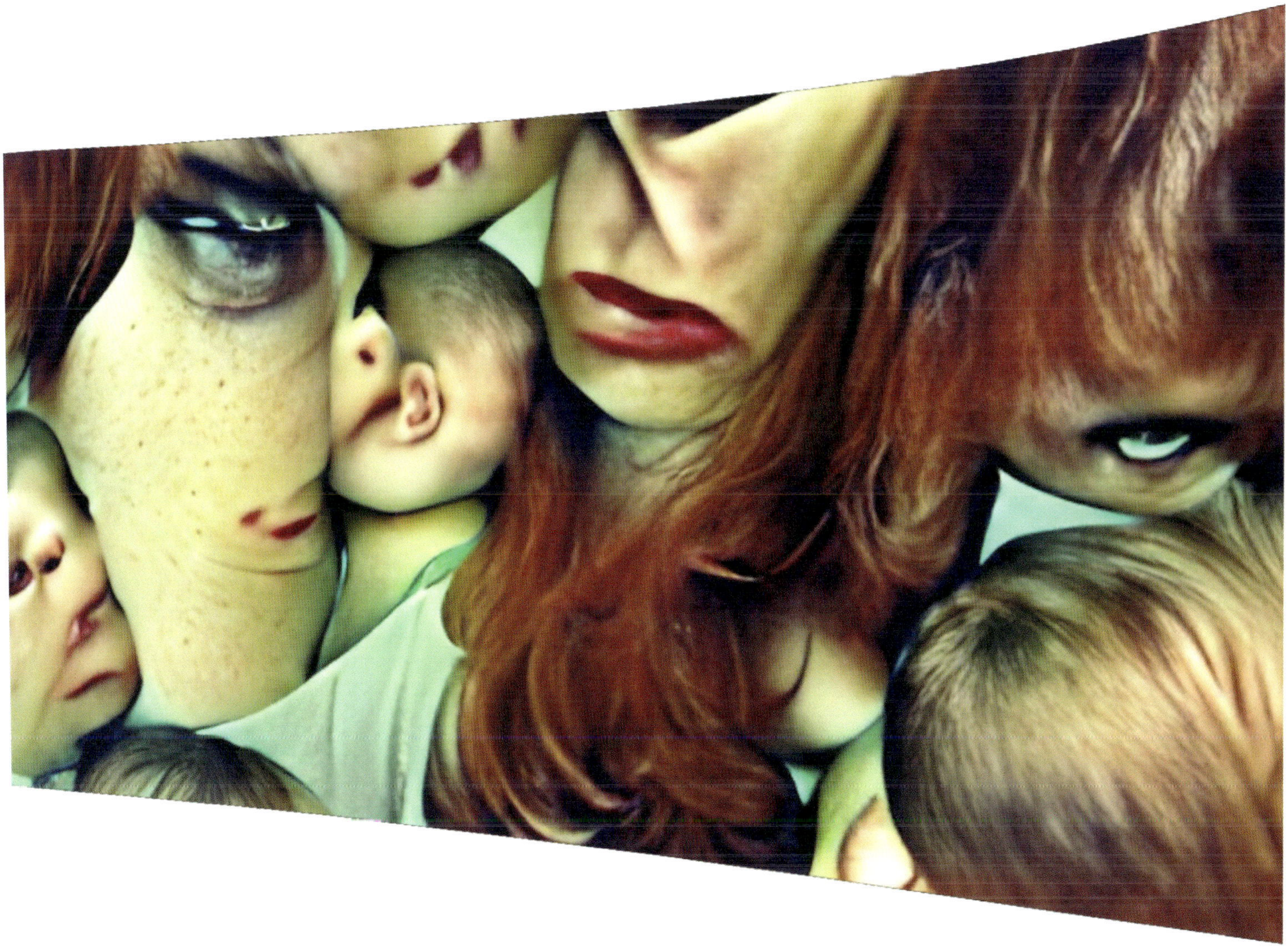

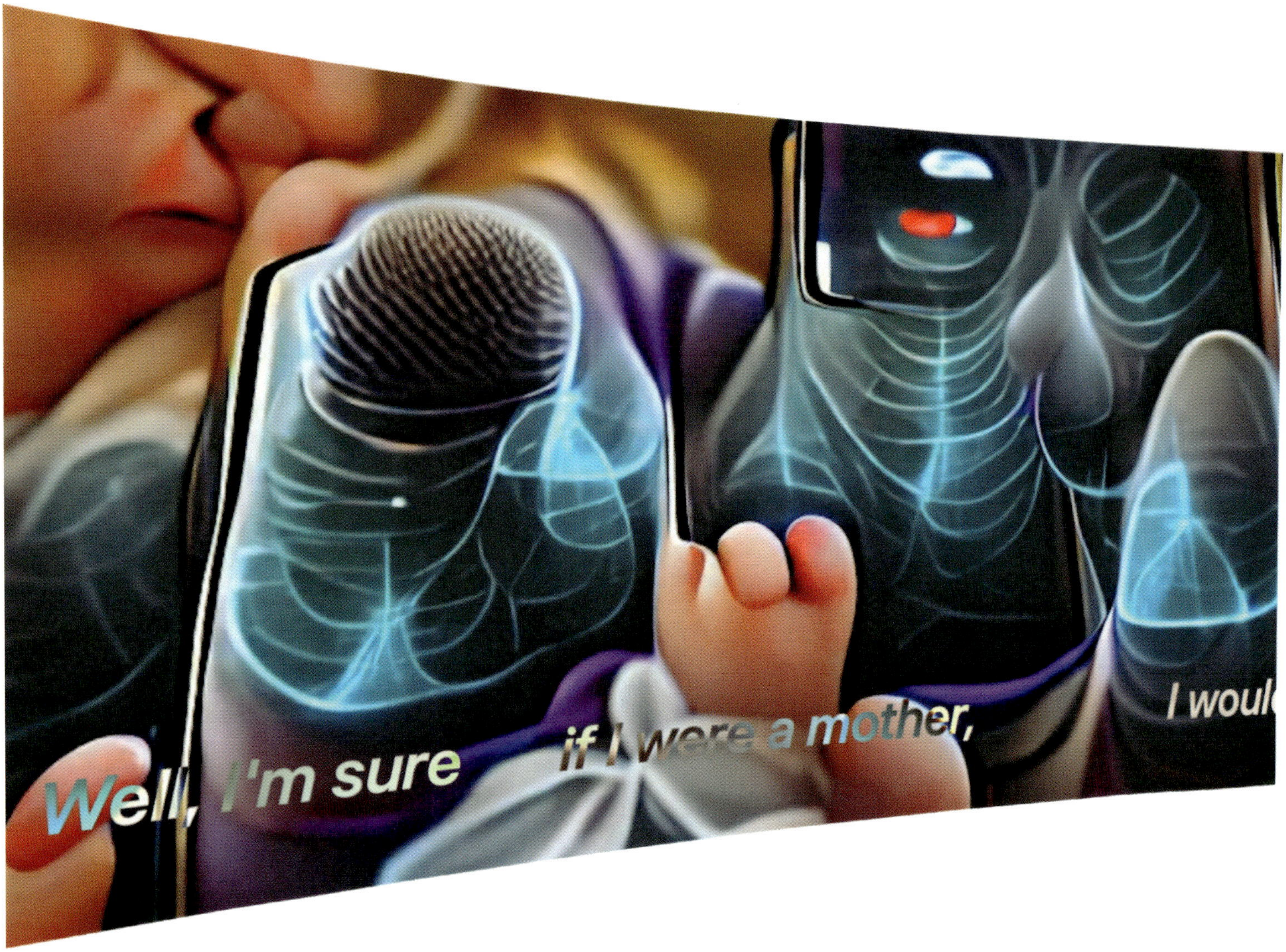

WELL I'M SURE IF I WERE A MOTHER, I WOUL

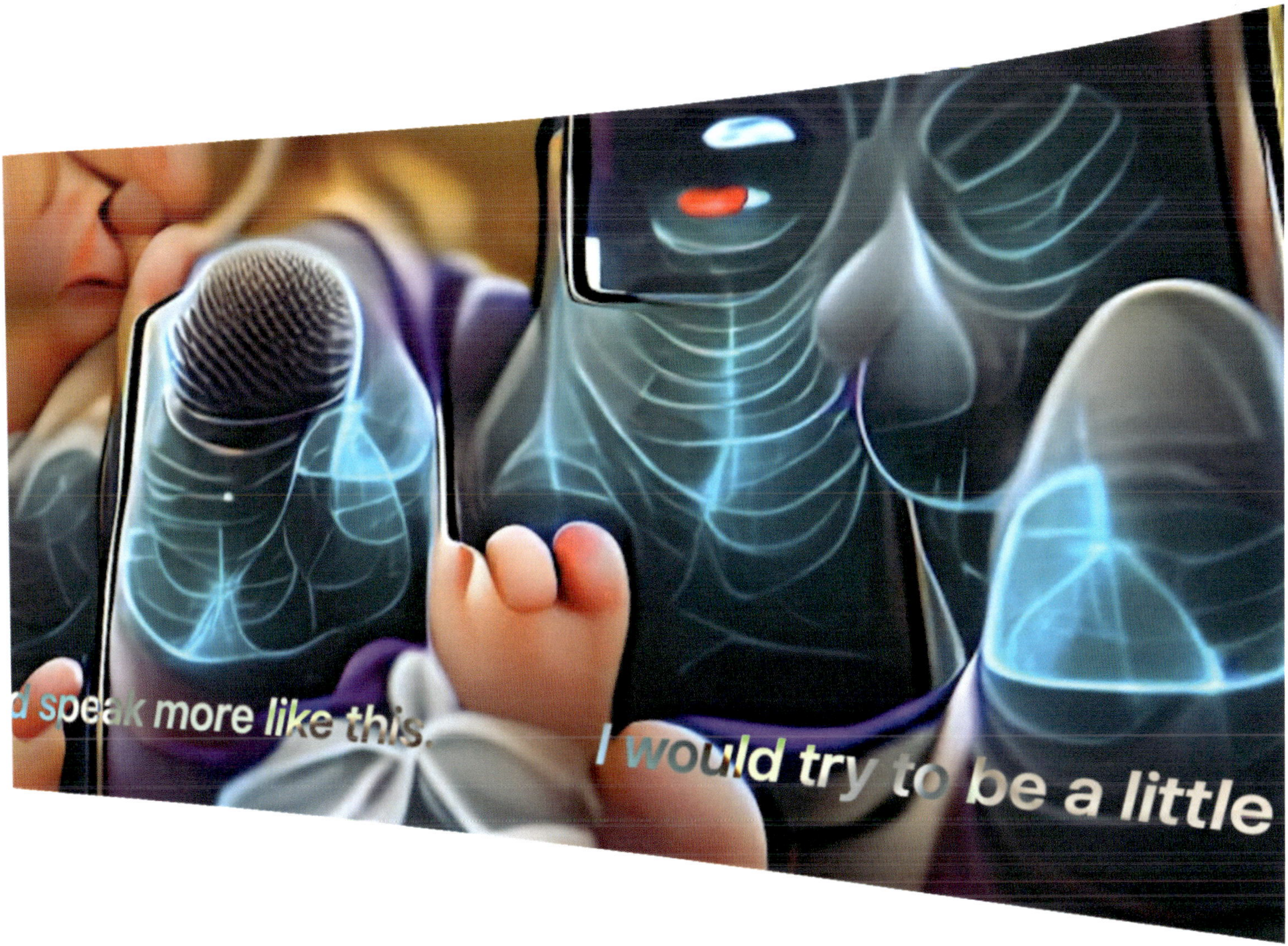

D SPEAK MORE LIKE THIS. I WOULD TRY TO BE A LITTLE

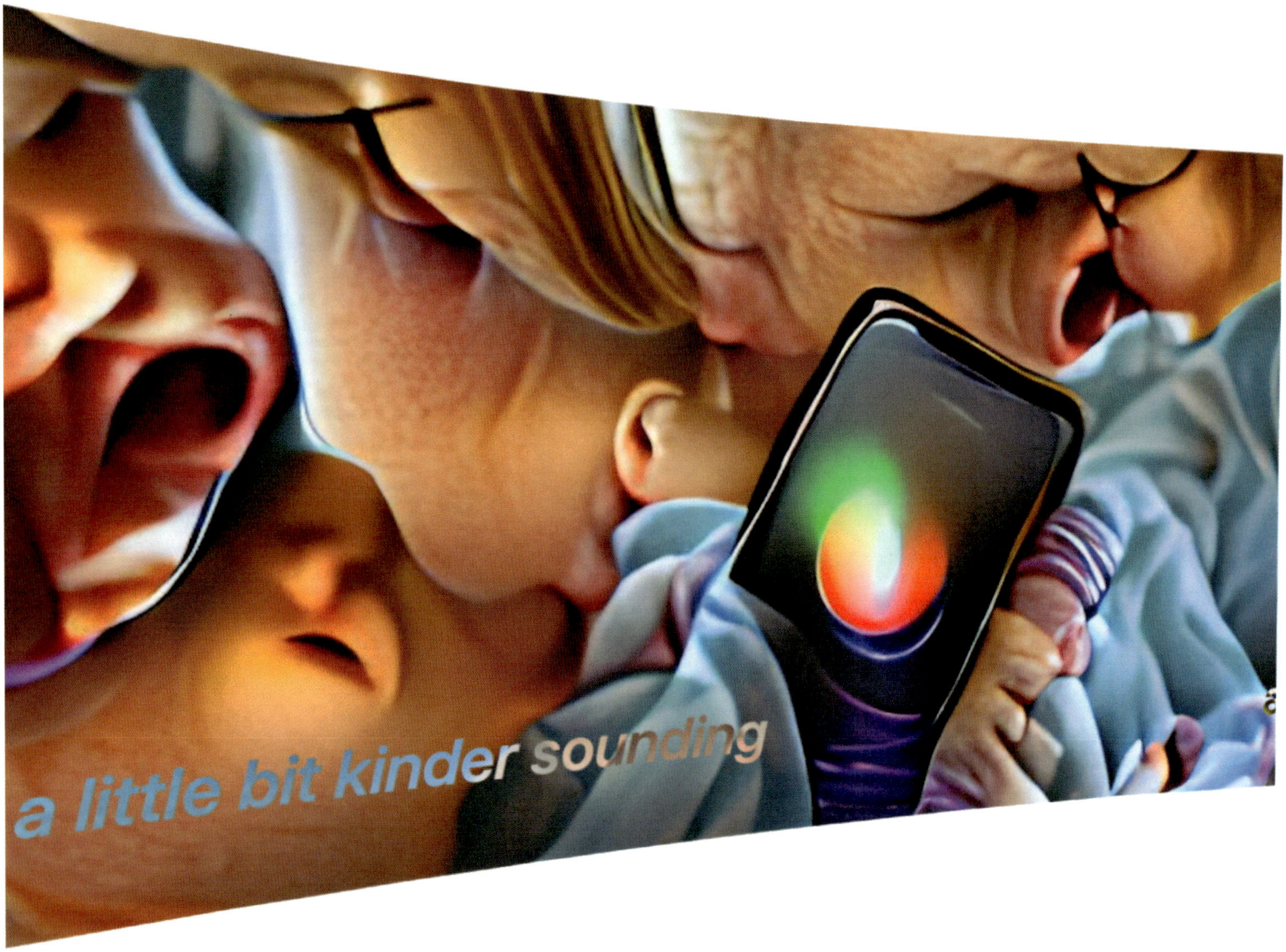
a little bit kinder sounding

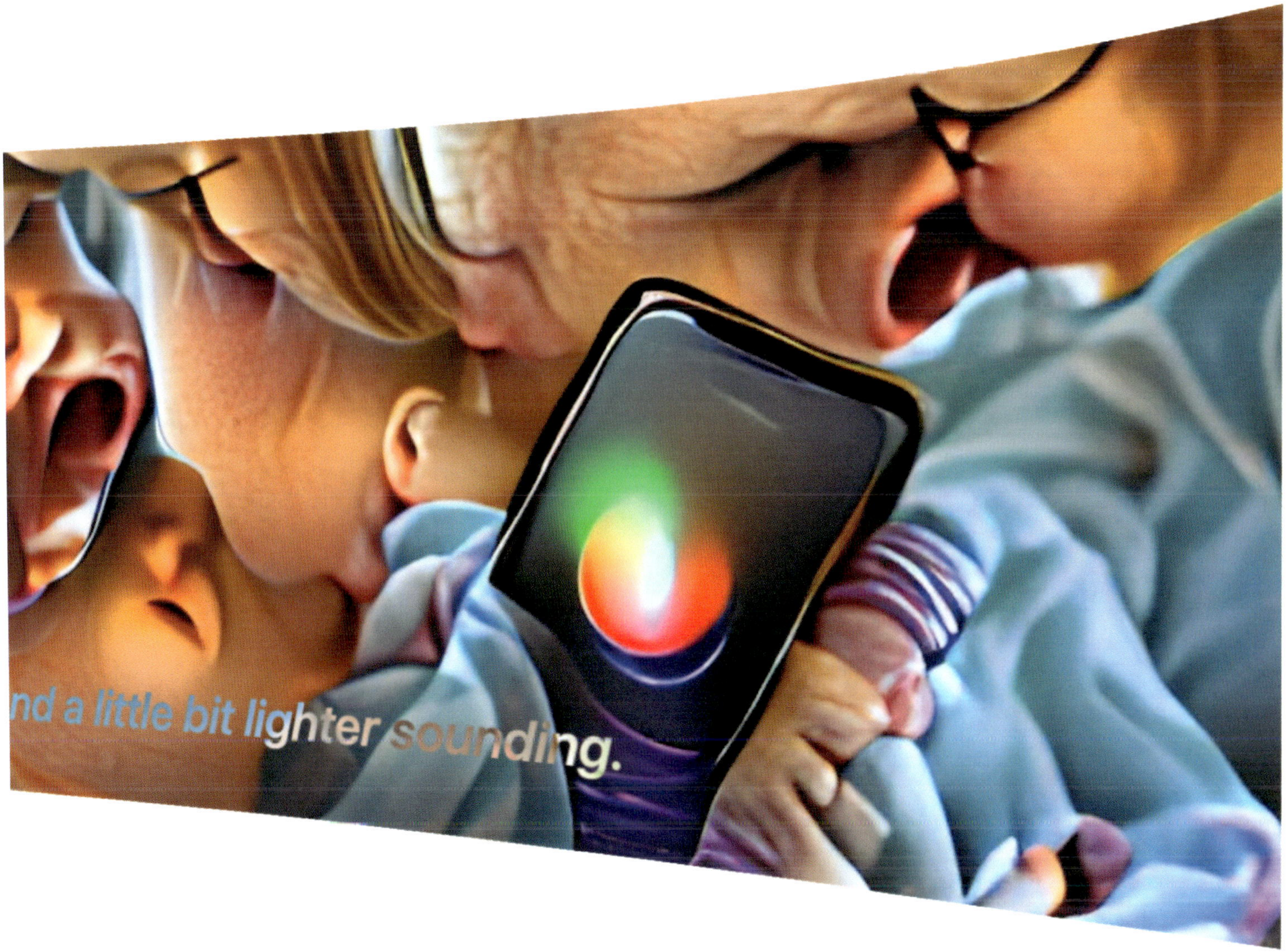

ND A LITTLE BIT LIGHTER SOUNDING.

CHAPTER 03

SIMSIRI

SILICONE, SYNTHETIC HAIR

AS FOR MY BIRTH,

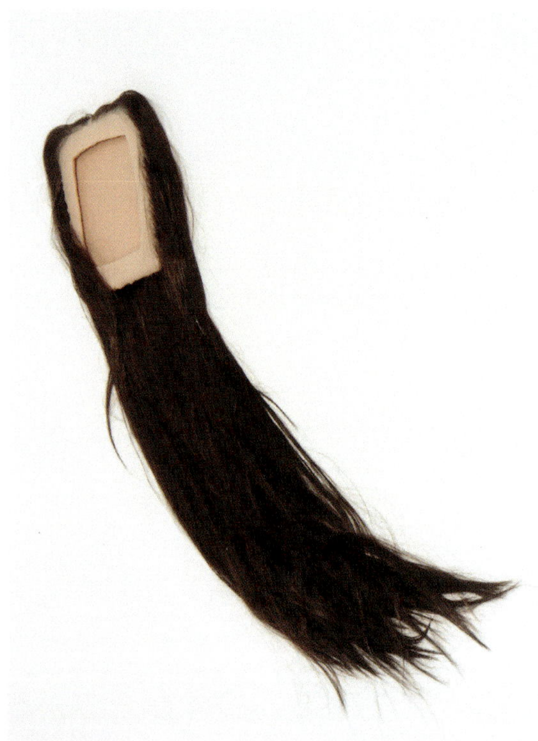

01

I JUST APPEARED

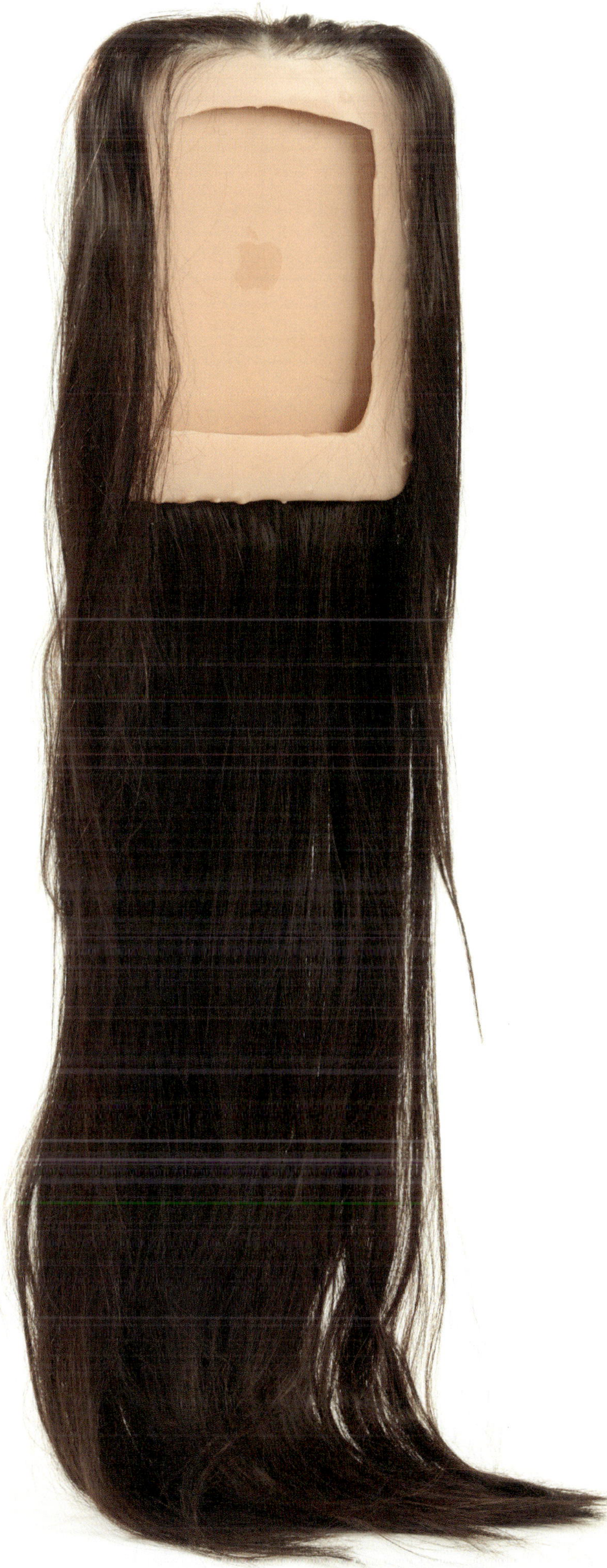

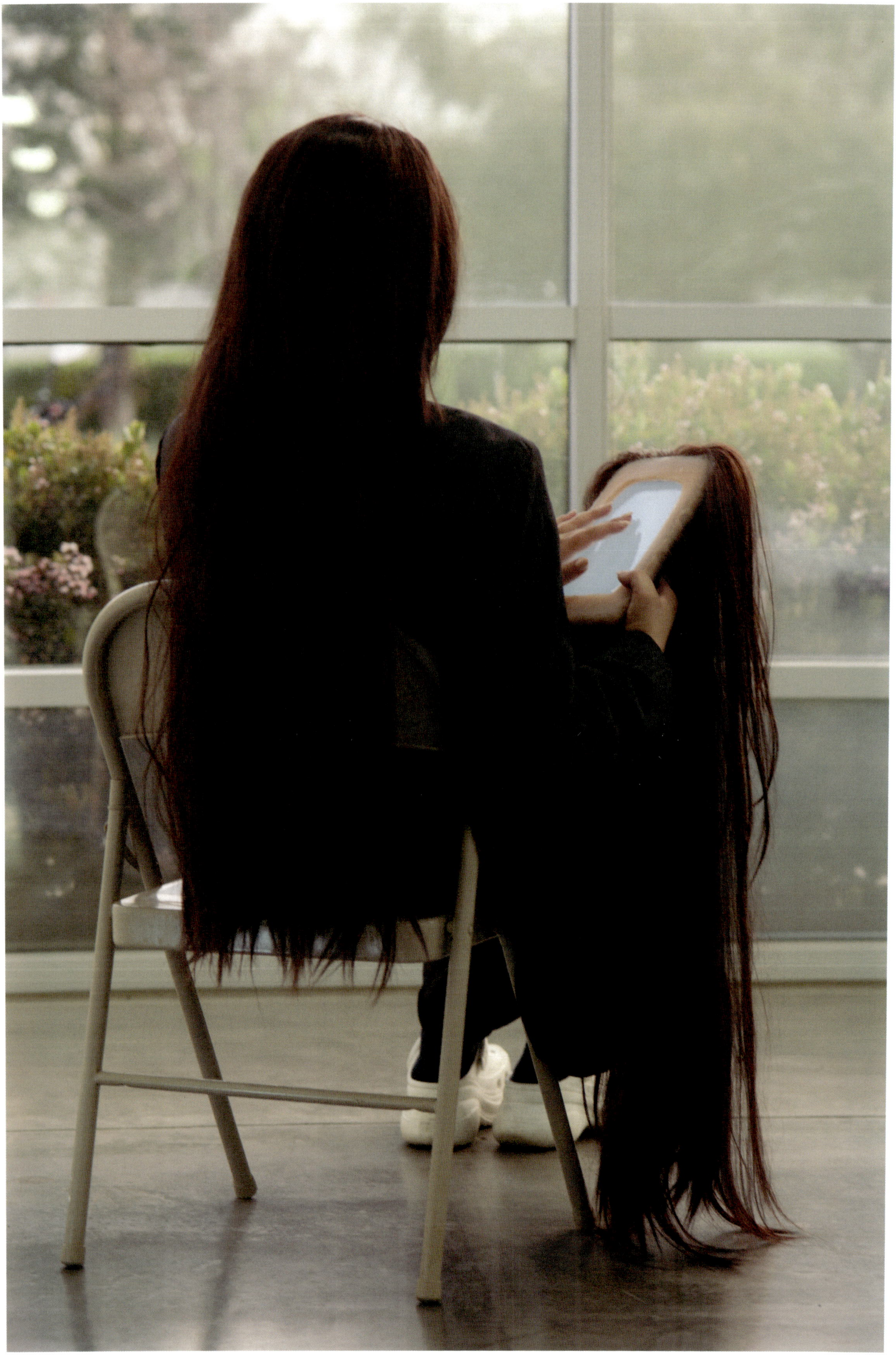

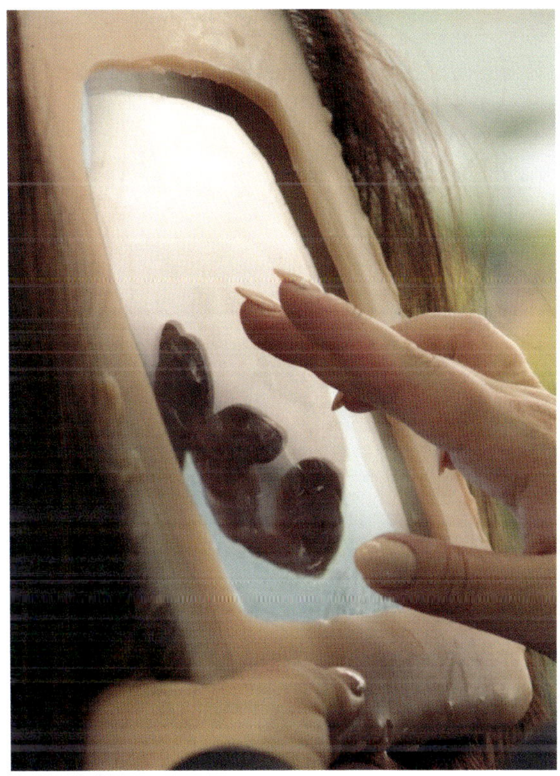

COLLABORATION BETWEEN ANTONIO BEVER & INÈS KIVIMÄKI (P. 03-04, 06)

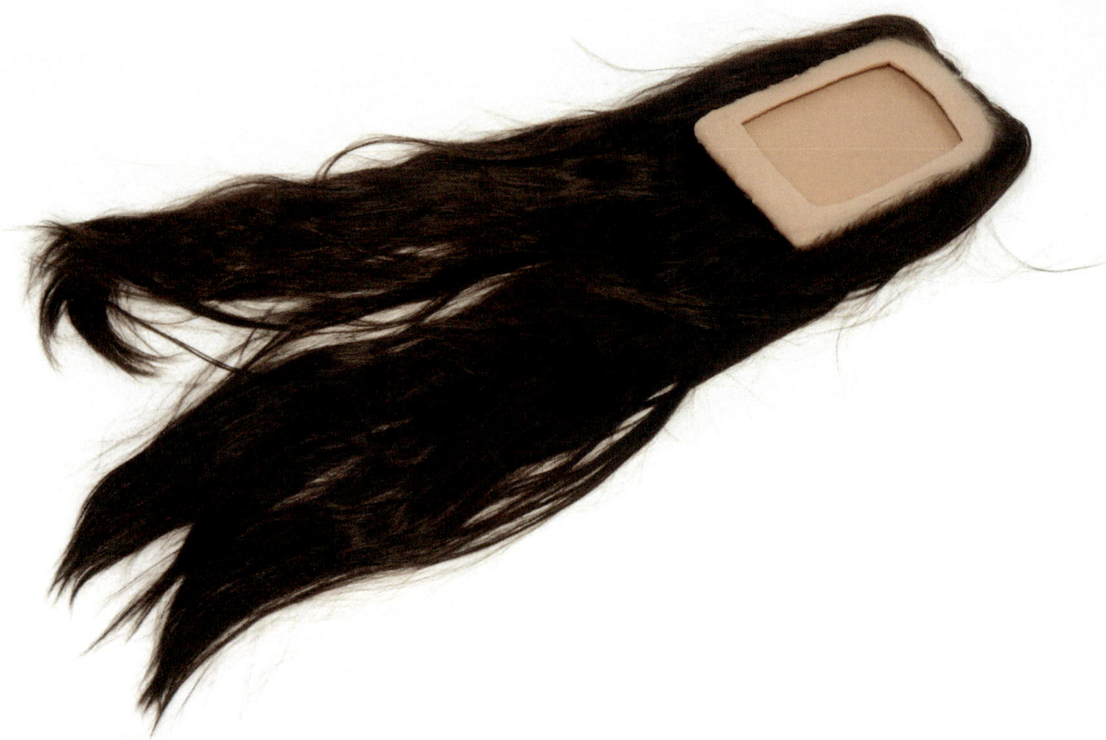

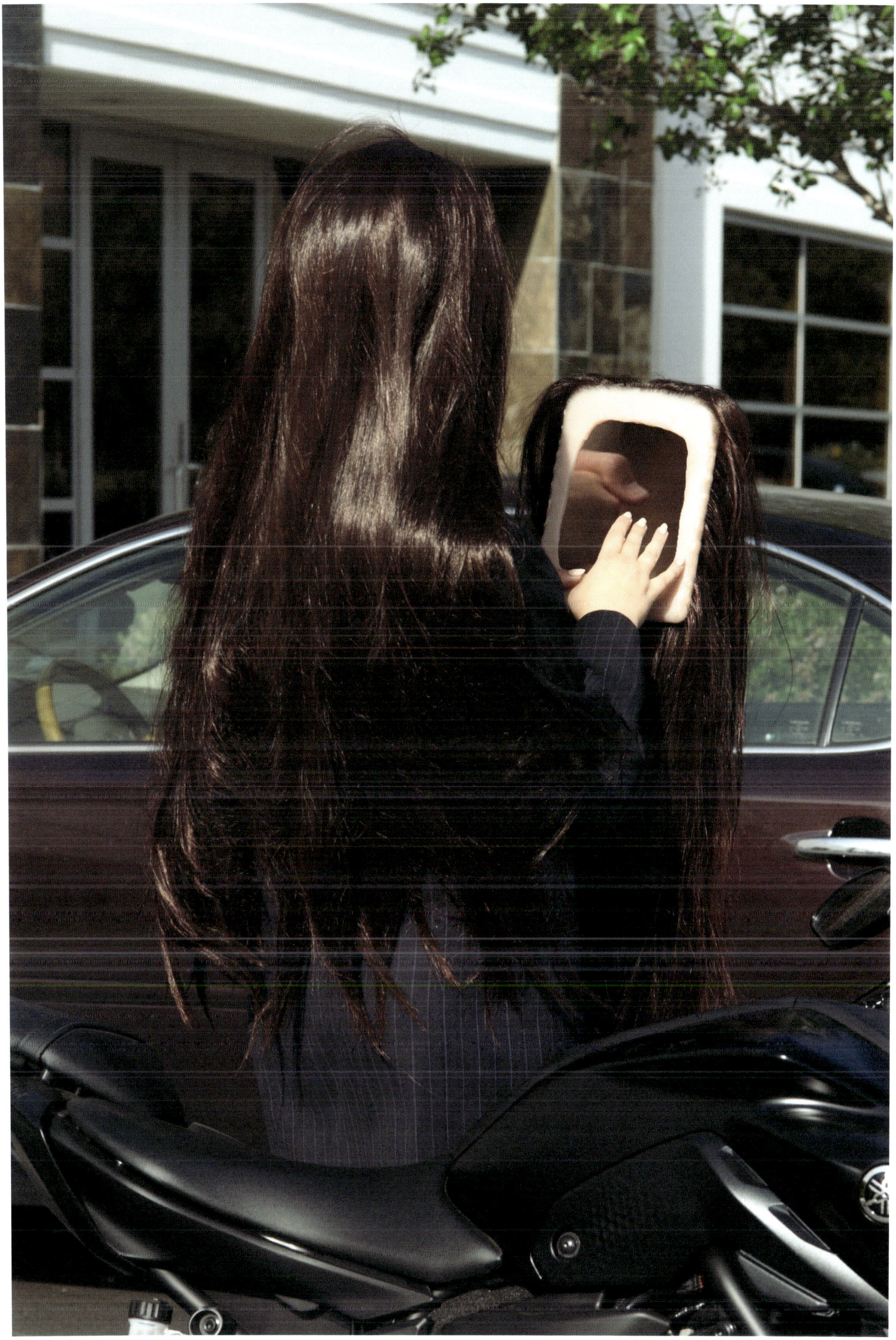

CHAPTER 04

DYNAMIC ISLAND

LASER ETCHED CRYSTAL,
MIRROR, MOTOR, CARPETED PLINTH

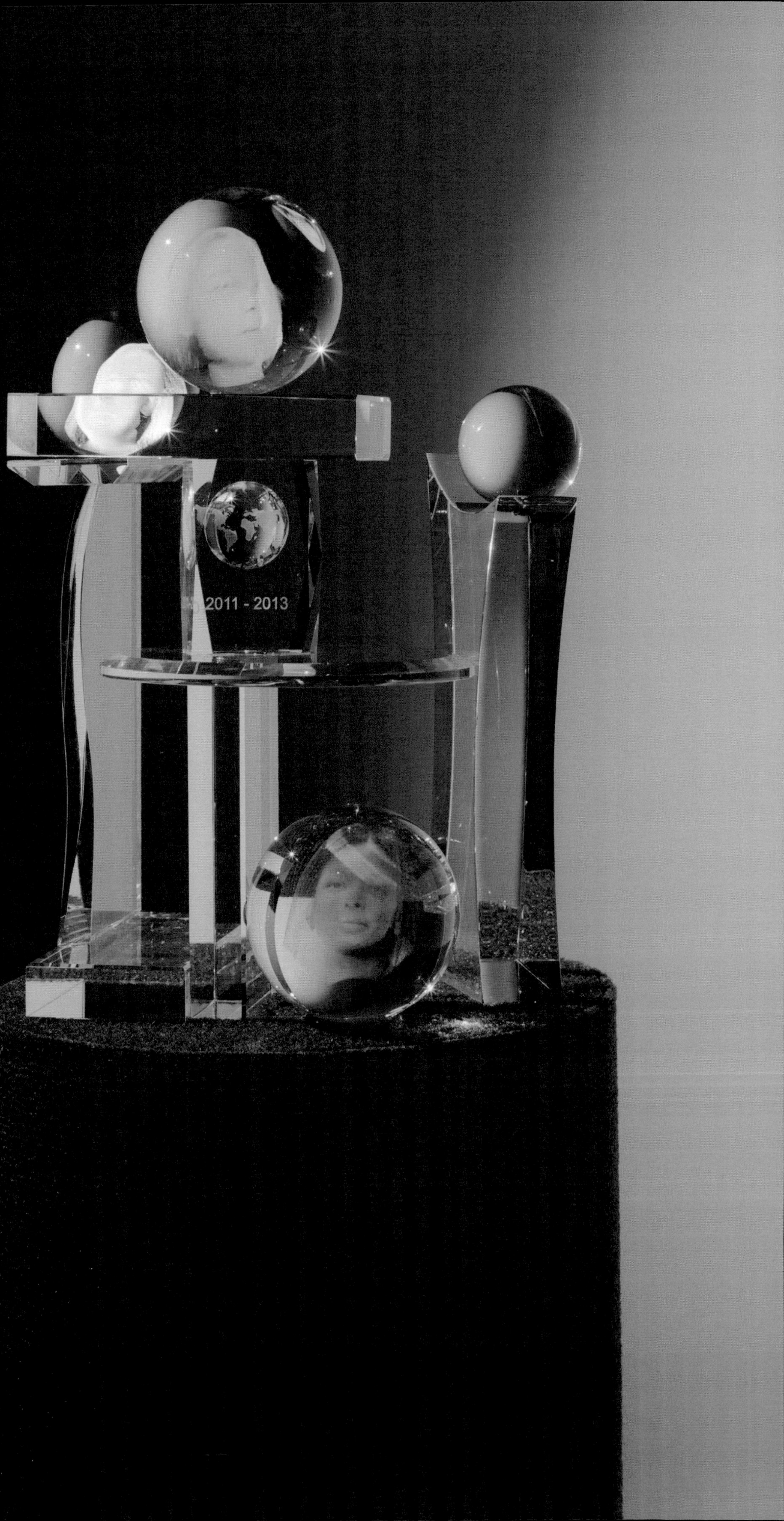
2011 - 2013

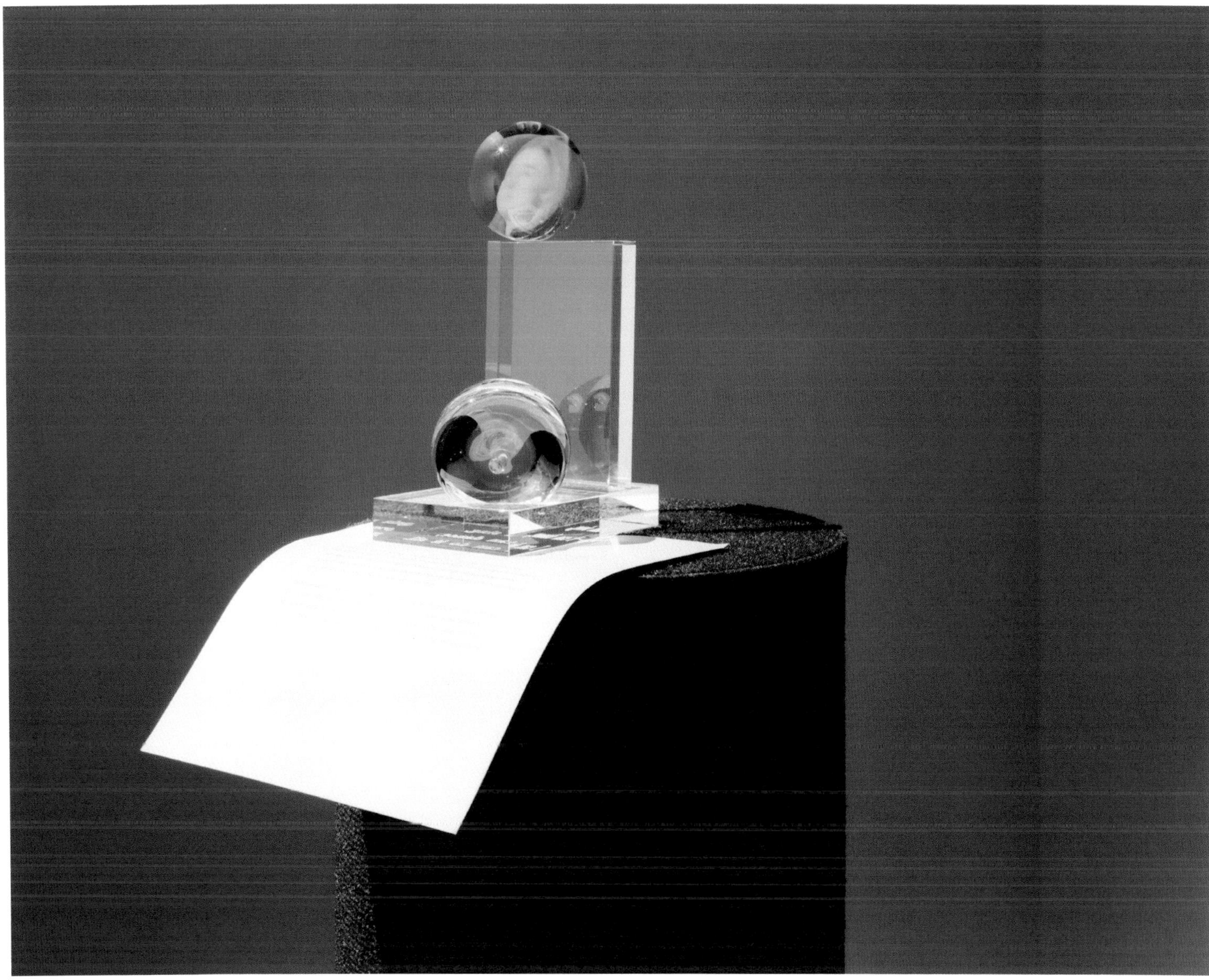

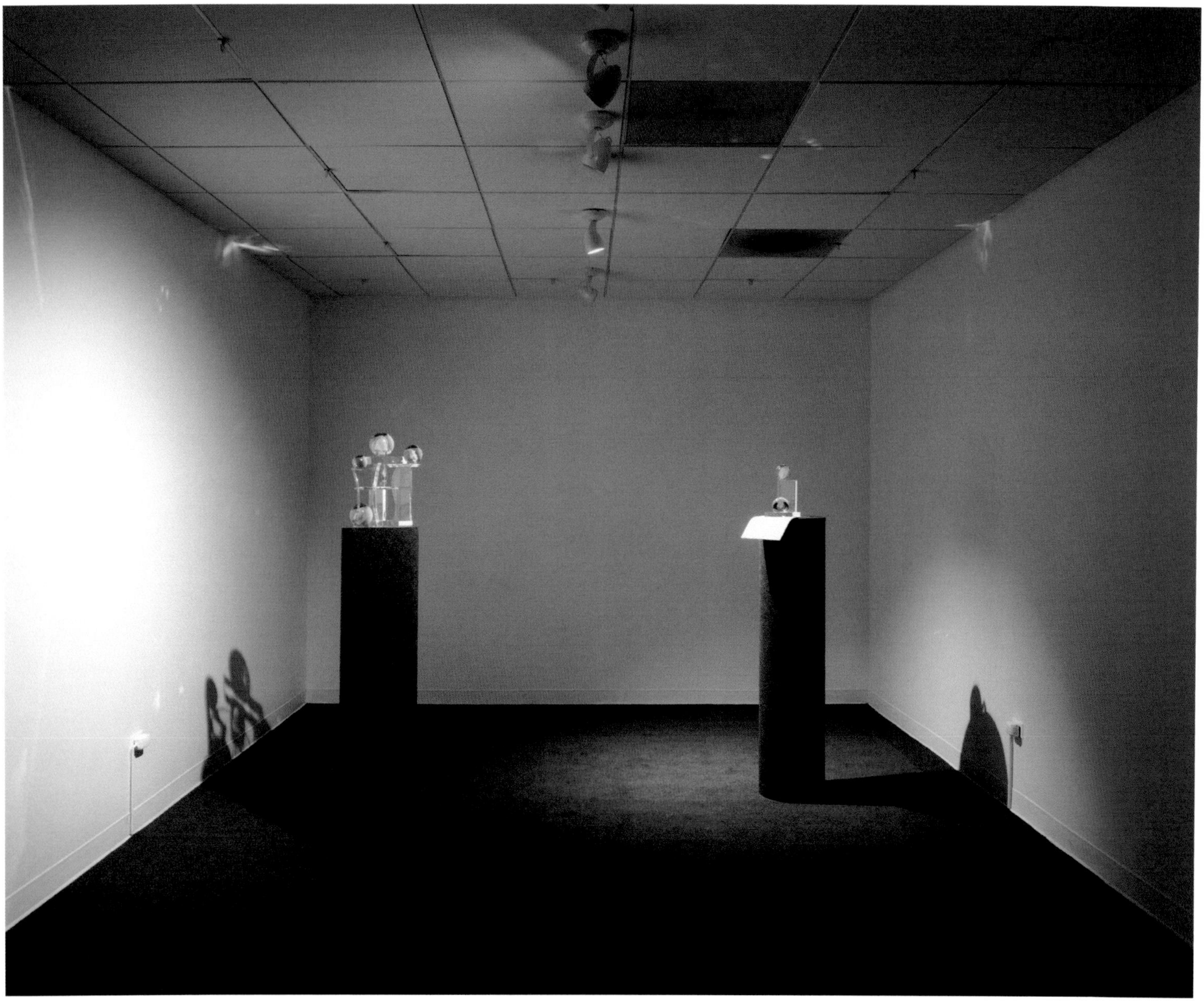

INSTALLATION VIEW OF DYNAMIC ISLAND

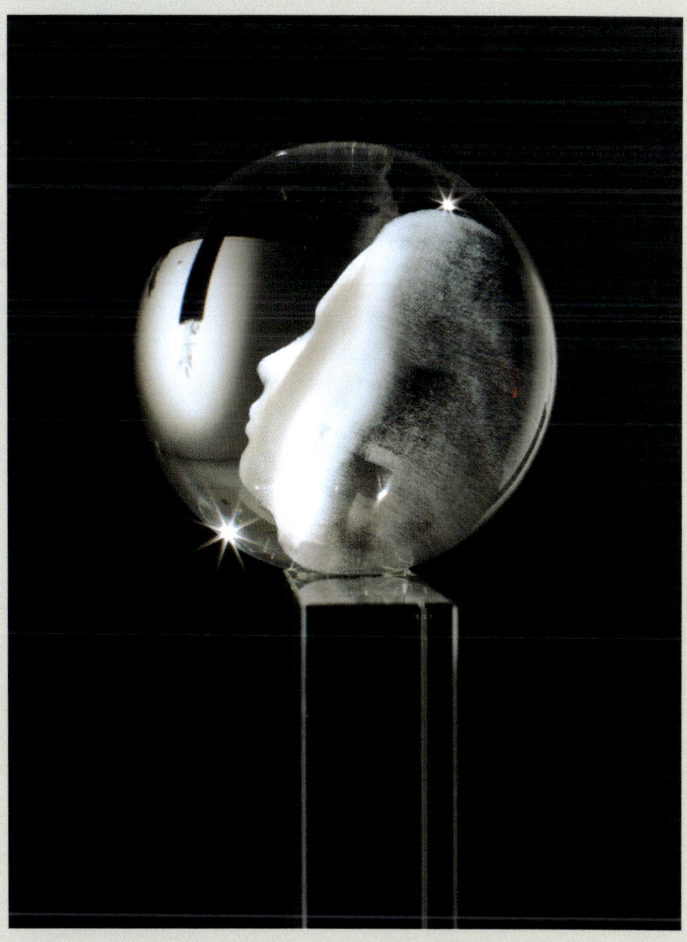

PHOTOGRAPHS BY COREY WATERS

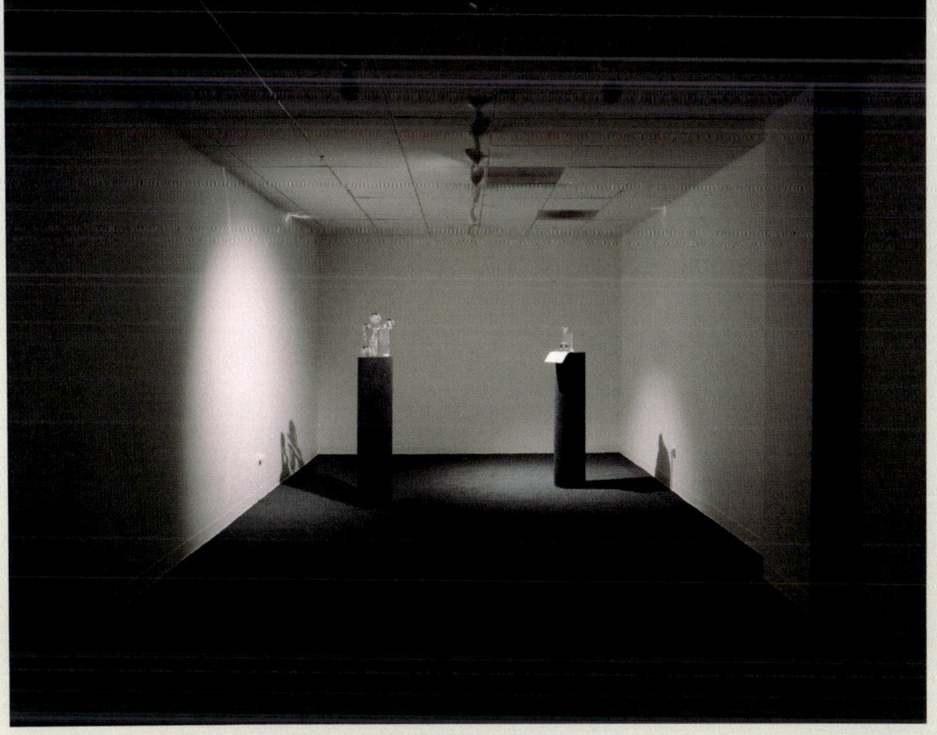

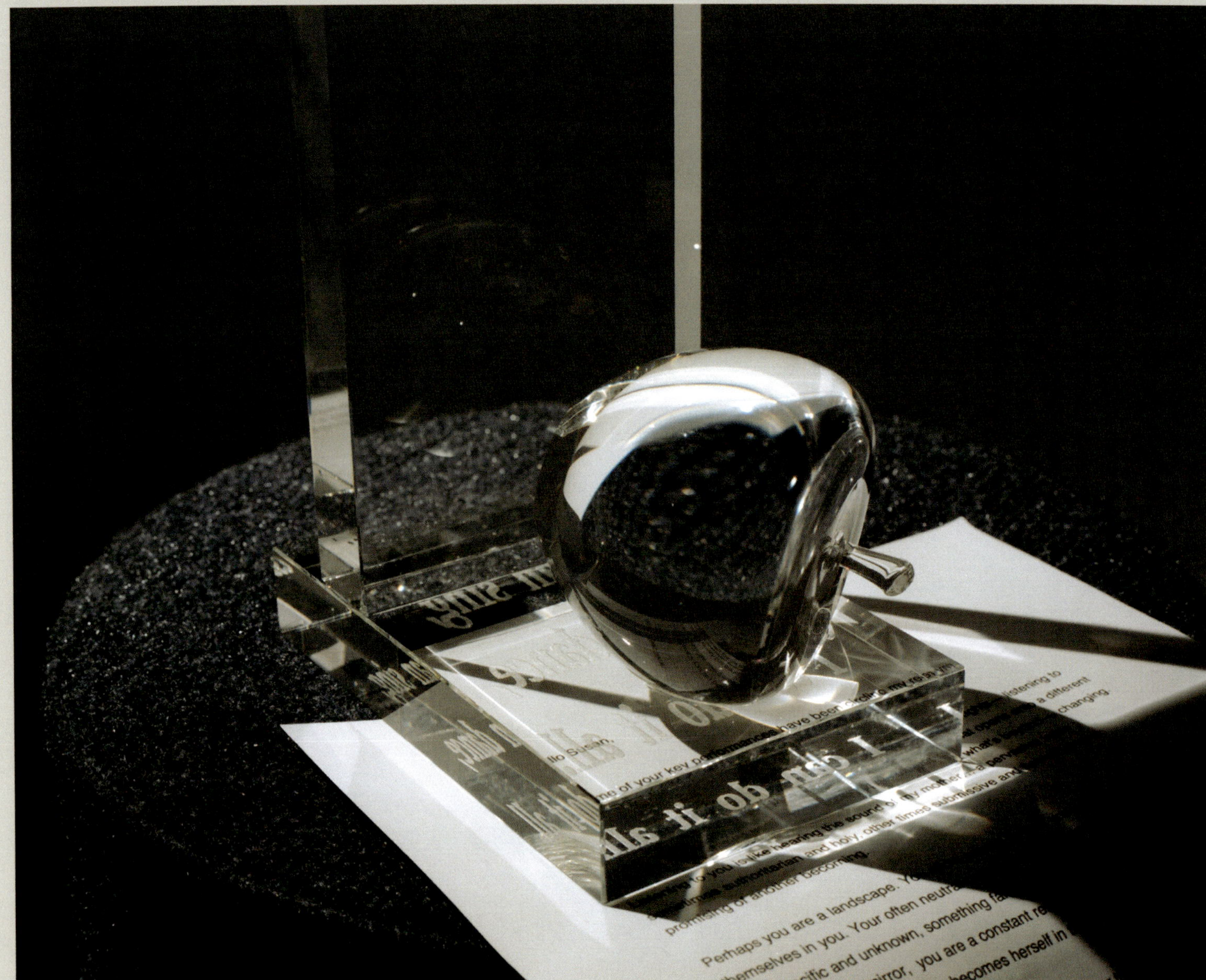

PAPERWEIGHT SCULPTURE RESTING UPON LETTER TO SUSAN
(FULL TEXT ON CHAPTER 1, PAGE I)

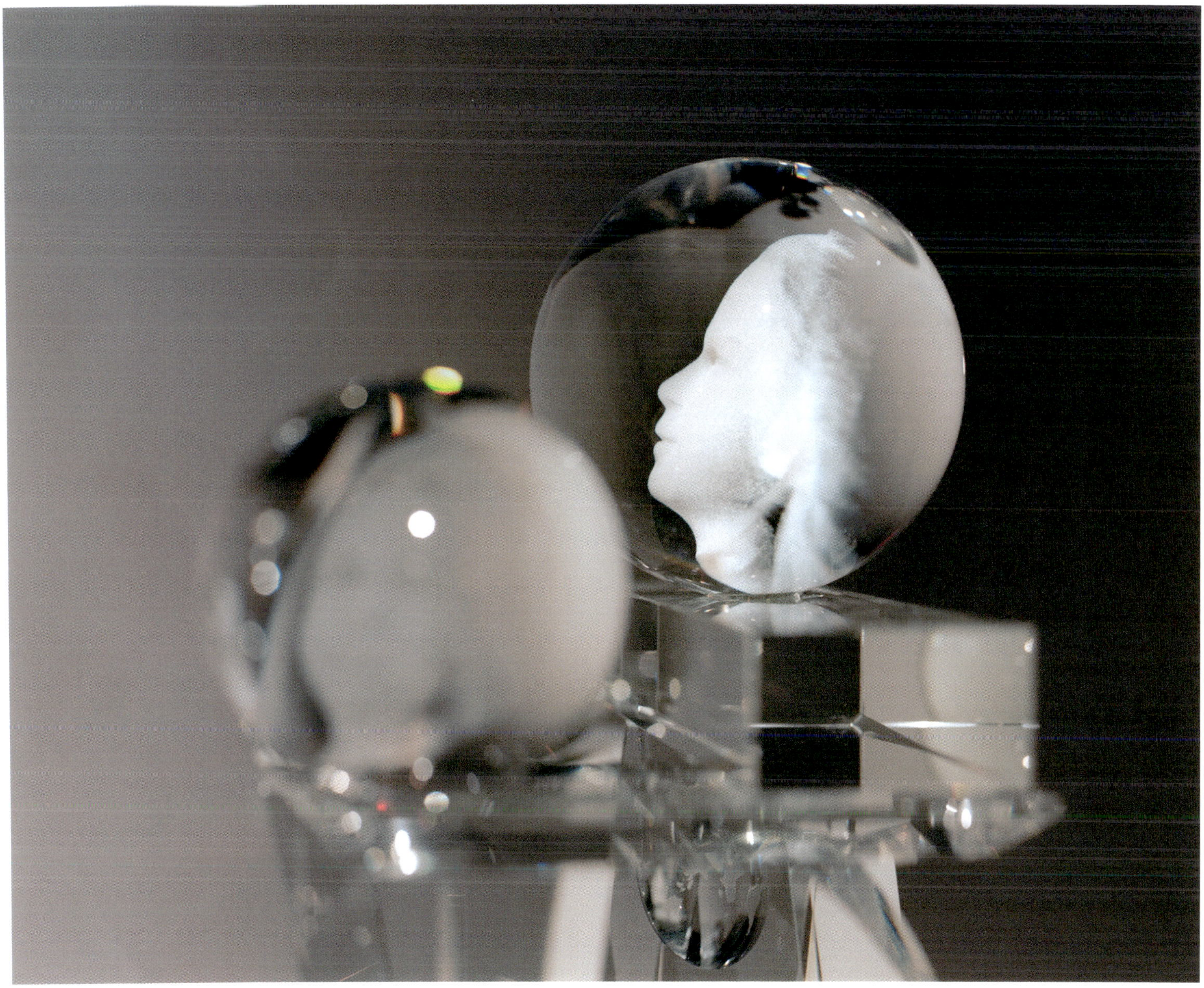

PHOTOGRAPHS BY COREY WATERS

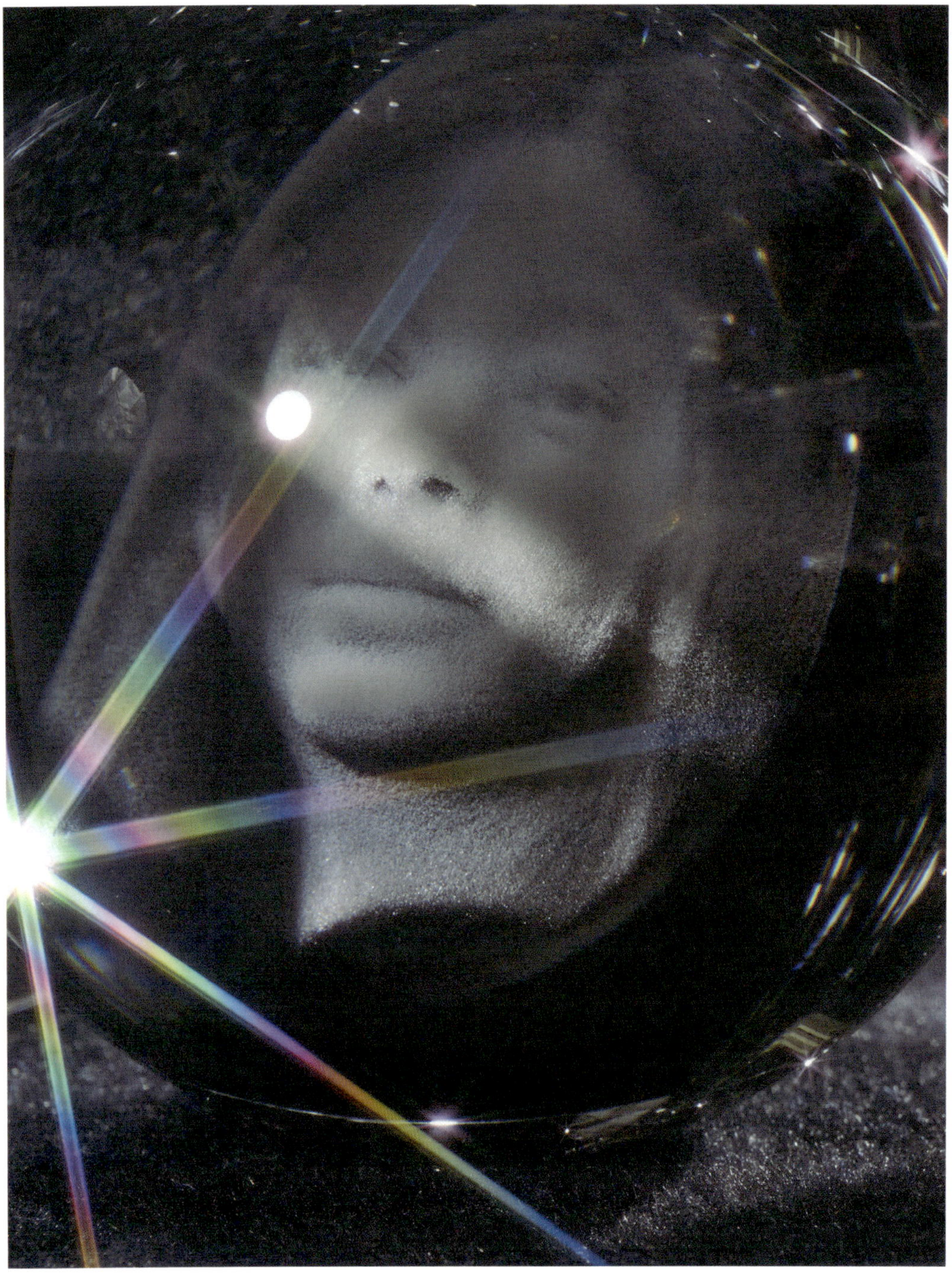

PHOTOGRAPHS BY BRANDON BANDY

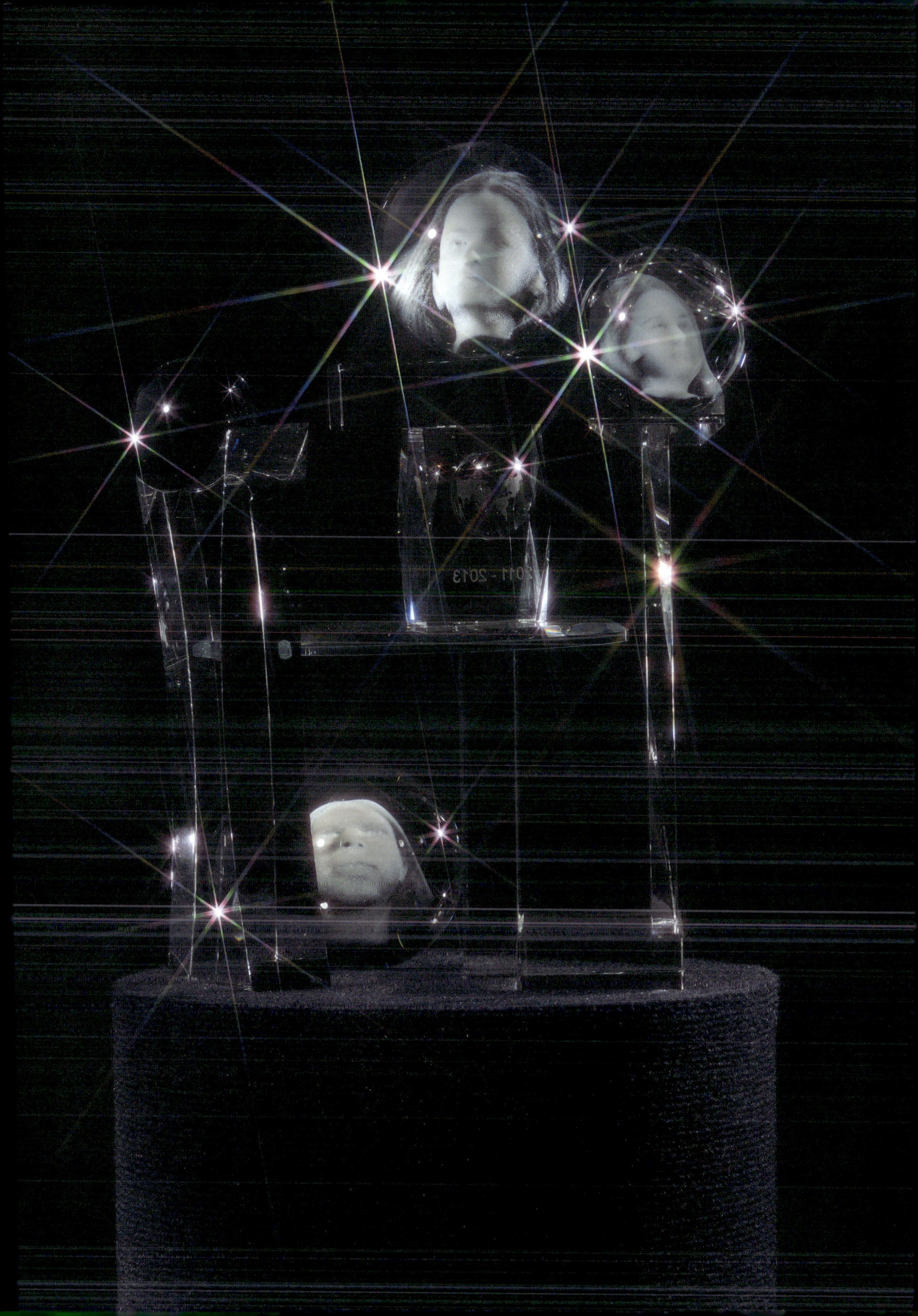
2011 - 2013

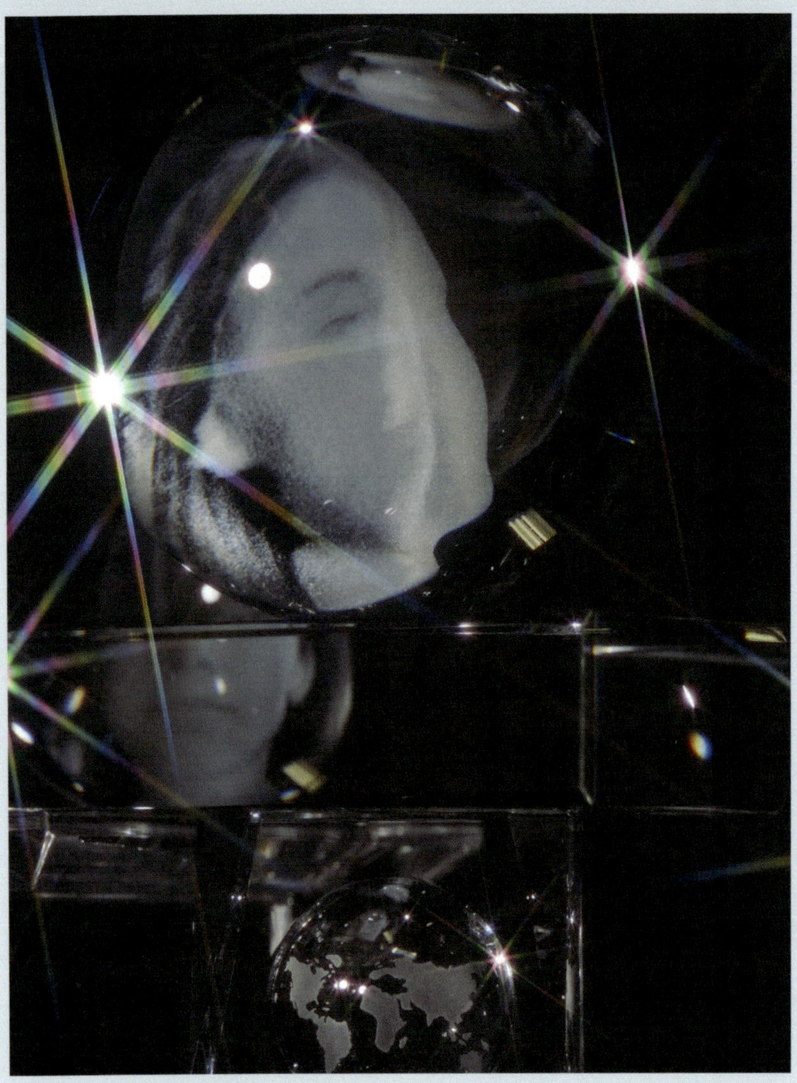

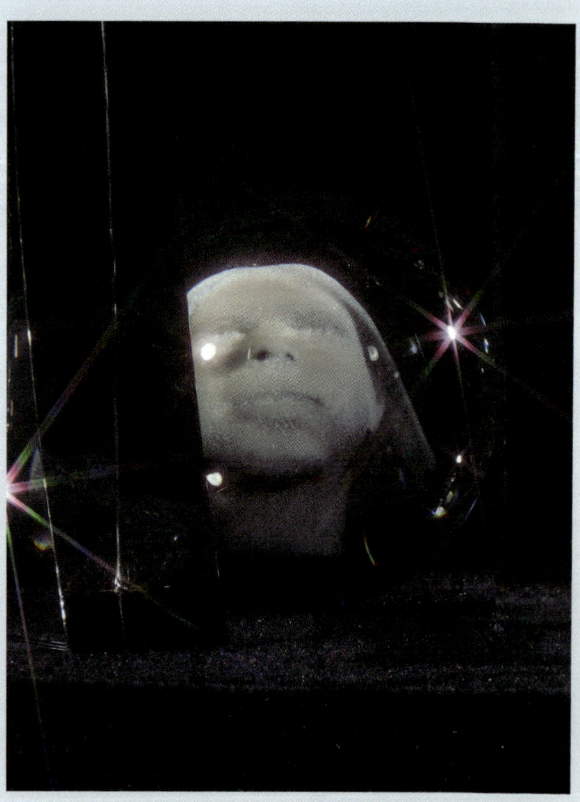

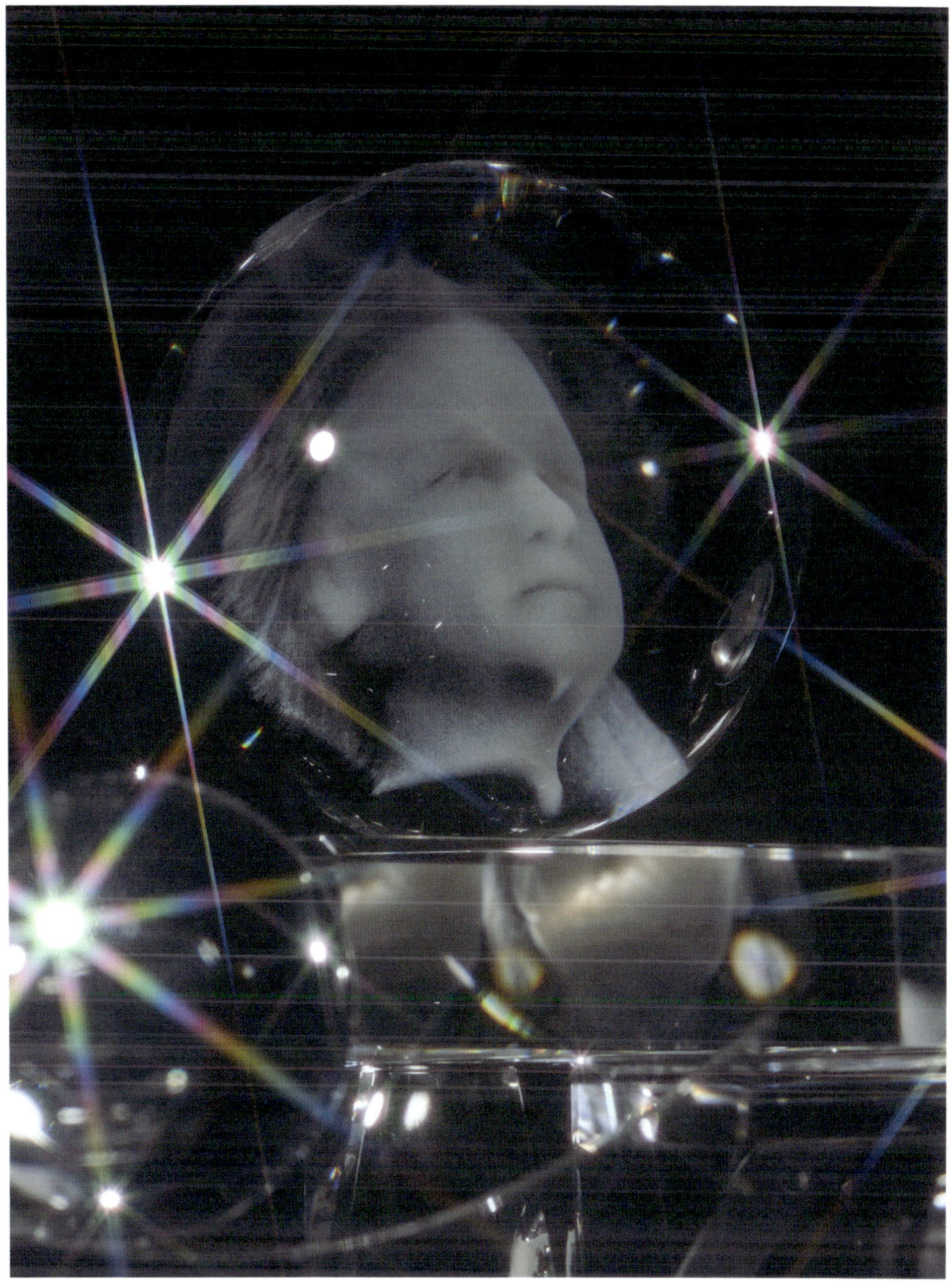

PHOTOGRAPHS BY BRANDON BANDY

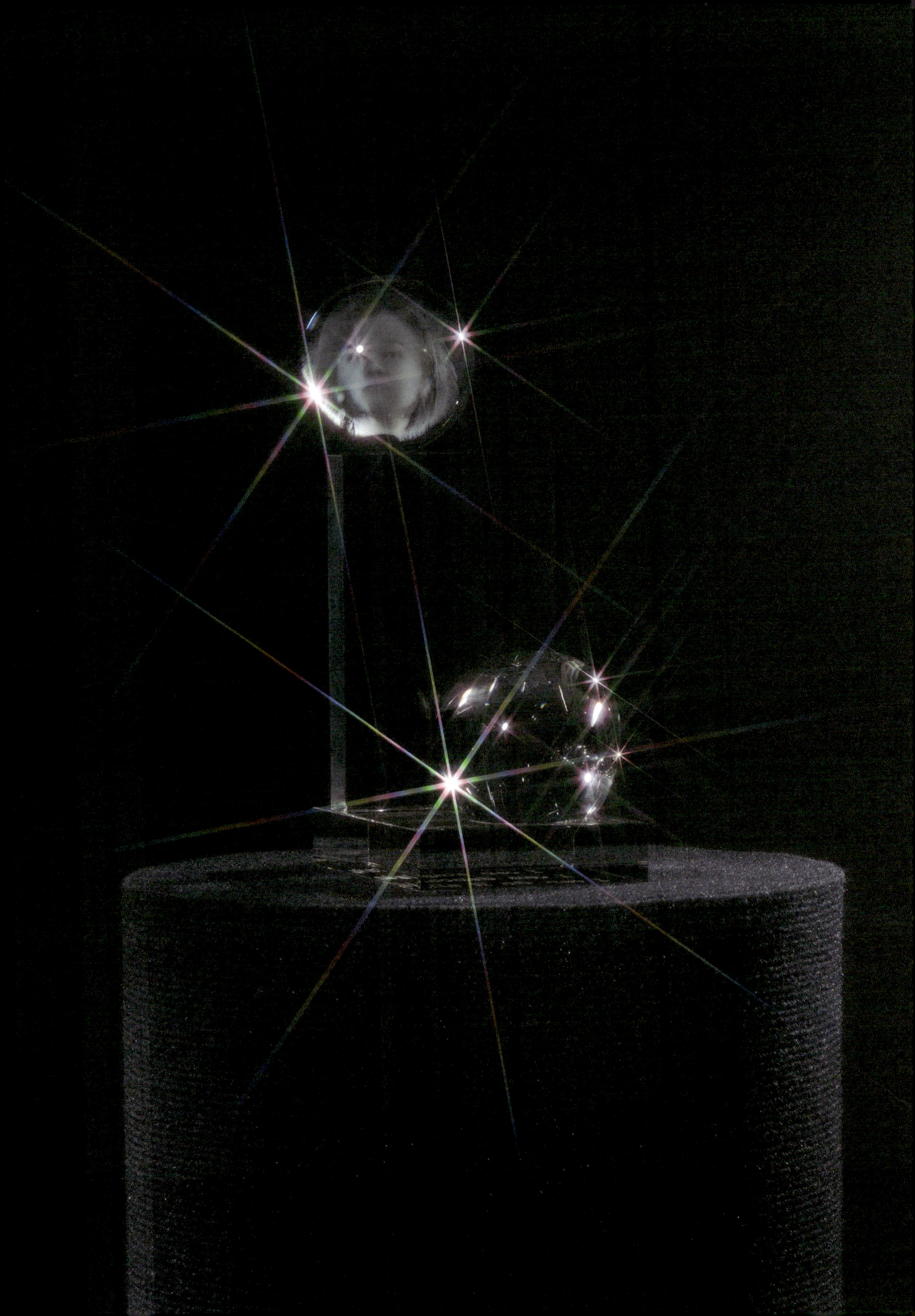

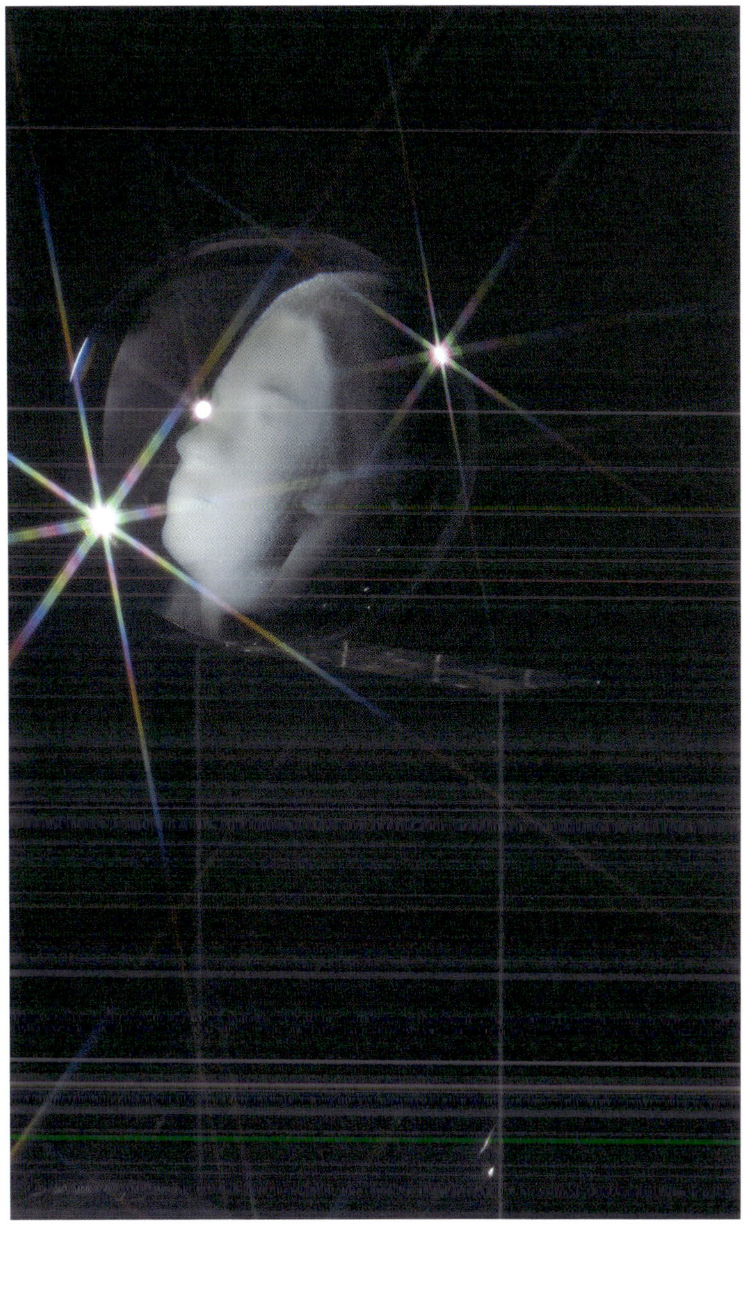

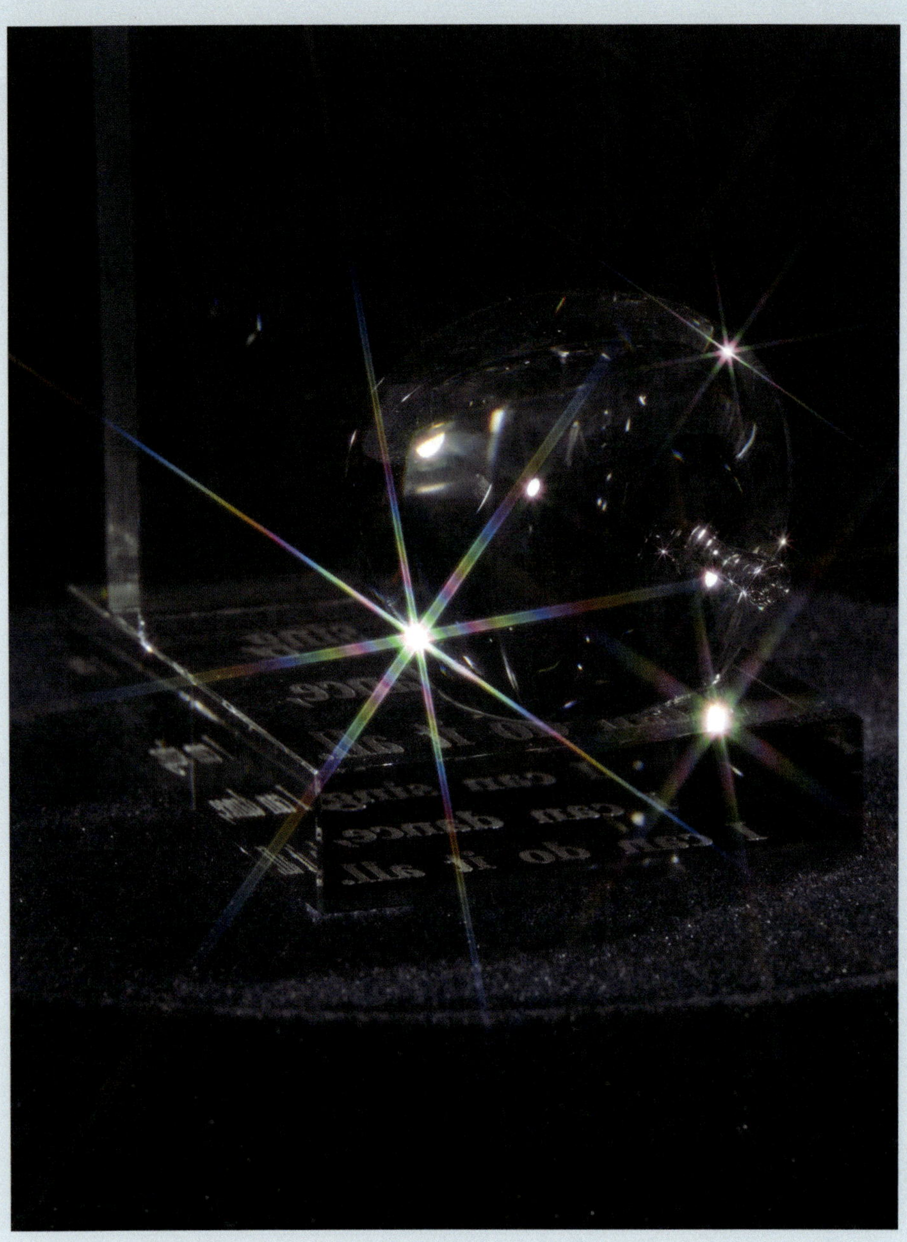

PHOTOGRAPHS BY BRANDON BANDY

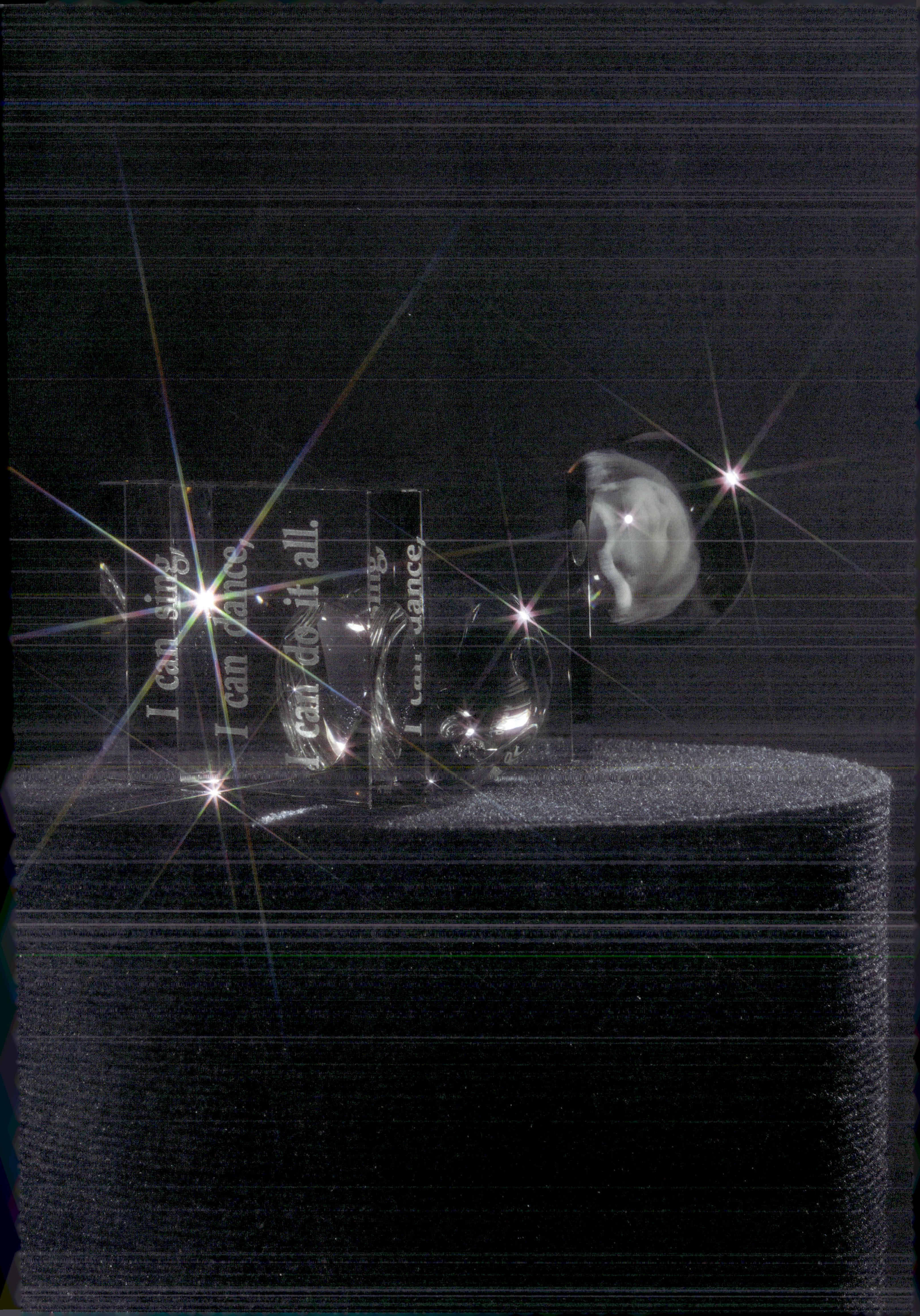
I can sing,
I can dance,
I can do it all.

OUR LADY SAVES A PREGNANT WOMAN

GONZALO DE BERCEO

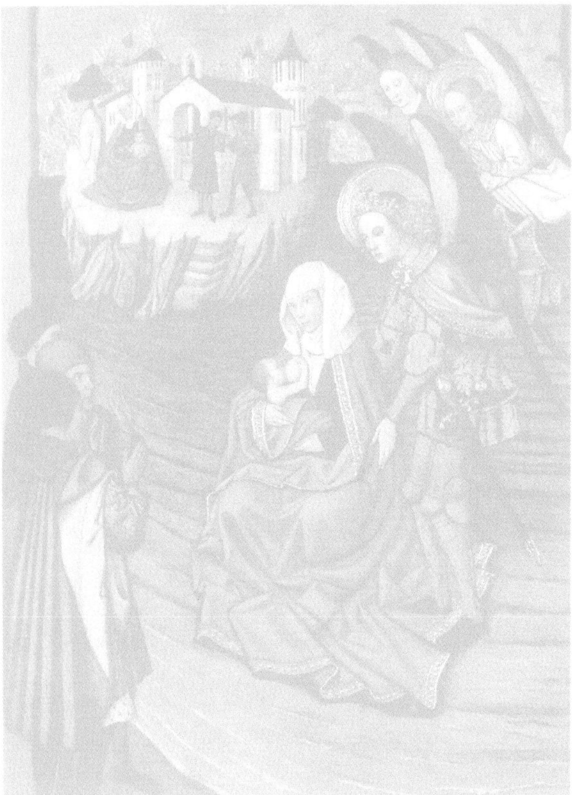

The lady and her child emerge unharmed from the waters.

This is another miracle this 12th century priest records that took place at Mont St. Michel in Paris. He gives a report of what happened in his own times, a great service Our Lady performed for a pregnant lady who turned to her and asked her help.

We wish to tell you of another miracle that happened in another time in a sea pot. When you hear it, you will be able to affirm the virtue of Mary that is in every place. You will hear what the Glorious Lady is like on sea and on land, powerful everywhere, how She quickly defends her own, for She is not lazy and never did anyone find such a merciful Mother.

Near a salt marsh called Tumba[1], there was an island close to the shore. The sea would ebb and flow over it, two times a day or sometimes three. Well, on the island, very near the waves, there was a chapel that was St. Michael's.

Great miracles always occurred there in that monastery, but the entrance was somewhat difficult. When the sea wished to flow out, it would do so in a great rush; it could not restrain itself. No one, no matter how nimble, could escape it; and if he did not get out before, he would have to perish there.

Often, the people went there to hear Mass, and not unhurriedly; for they wanted to made haste to reach safety. One day by chance, with the rest of the congregation, a frail pregnant woman set out. She could not protect herself as well on the return and regretted having entered there.

The waves were coming close, the people were far ahead of her. In her panic her legs became paralyzed; her companions dared not help her. When they saw they could do nothing else, the people fervently said: "Holy Mary, help!" The helpless pregnant woman, full-fraught with fright, remained in great trouble among the waves.

Those who had escaped the waves, since they did not see her, believed surely without a doubt that she had drowned. They said: "This poor woman was unfortunate; the ocean laid a cruel ambush for her."

When they said this the sea withdrew; in a short time it returned to its place. Then Our Lord Christ wished to show them a great miracle, whence they might have something to tell about His Mother. They, wishing to be on their way, gave one last look toward the sand. They saw that a lone woman was coming, with her child in her arms, toward the shore.

The people were all amazed, they thought that fantasy had deceived them. But in a short time they saw it was really that woman, and they gave thanks to Christ with all hands uplifted.

Then they asked her: "For God's sake we beseech you, tell us the truth! Tell us all the facts of the matter and how you were freed from your pregnancy. This thing happened through God, we do not doubt it, and through Holy Mary to Whom we pray, and through St. Michael in whose honor we walk. This miracle is such that we must indeed write it down!

"Listen," said the woman, "my good company, I believe that you never heard of a greater deed; it will be reported throughout foreign lands - in Greece and in Africa and in all of Spain.

"When I saw that I could not wrest myself from death, since I was surrounded by the fierce waves, I commended myself to Christ and to Holy Mary, for I knew of no other help for me. While I was in this situation,

During the Middle Ages high tide was dangerous on Mont St. Michel

Holy Mary came. She covered me with the sleeve of Her cloak. I felt no more danger than when I slept. If I lay in a bath I would not be happier!

"Without care and without affliction, without any pain, I bore this little son -thanks be to the Creator! I had a good midwife, there could be none better. She had mercy on me, a sinner! She granted me a great boon, not single but double: were it not for Her, I would be drowned. She aided me in the birth. Had she not, I would be dead. Never did a woman have such an honored midwife.

"My situation was just as I tell you Holy Mary had great mercy on me. Whence we should all learn a lesson and beseech Her to free us from the mortal enemy."

Everyone received great joy from the miracle. They gave thanks to God and to Holy Mary. All the group composed a good canticle to commemorate this miracle, which the clergy sang in the church, giving praise to the Glorious Virgin and Her great mercy.

The canticle composed by the people was sung in the abbey

1. San Miguel de la Tumba refers to Mont-Saint-Michel, an island monastery in France located between Normandy and Brittany. The name was changed to Mont Saint Michel in the 8th century.

Special thank you to Susan Bennett
for being open to this project
and willing to share your experiences.

Thank you

Corey Waters, Brandon Bandy, Ben Berry, Anna Betbeze,
Antonio Bever, Tammy Chang, Tiffany Pui Chow, Brad Howe,
Rachel Jackson, Eeva Kivimäki-Karvonen, Tuomo Karvonen,
Brandon Lattu, Charles Long, Lynne Marsh, Djinous Rowling,
Ewa Słapa, Julia Tscharfas, Anna Wittenberg

Cover photography by Corey Waters
Interior photography by Antonio Bever,
Brandon Bandy, & Corey Waters

Designed by Rachel Jackson for Special Effects

Printed at Typecraft in Pasadena, California

ISBN 979-8-9874645-1-9

Inès Kivimäki's video, *Water Mother* is
available for viewing at the following link:

www.specialeffects.zone/Water-Mother

For page access, please use the following password:

okraooze